Common Bonds
Reflections of A Cancer Doctor

by

E. Roy Berger, M.D.

with Linda A. Mittiga

Previously Published by Health Education Literary Publisher
P.O. Box 948
Westbury, NY 11590

Health Education Literary Publisher is an affiliate of the National Center to Save
Our Schools

Library of Congress Cataloging-in-Publication Data
Berger, E. Roy, 1944-
    Common Bonds: reflections of a cancer doctor/by E. Roy Berger, with
    Linda A. Mittiga.
    p. cm.
    ISBN 1-883257-07-7: 24.95
    1. Berger, E. Roy, 1944-   2. Oncologist — United States —
    Biography. I. Mittiga, Linda A. II. Title.
    (DNLM: Neoplasms—therapy—personal narratives.
    2. Neoplasms—therapy—case studies. 3. Physician-Patient
    Relations—personal narratives.
    QZ 266 B496e1995)   RC279.6.B47A3—dc20
    (b)
    DNLM/DLC
    for Library of Congress   95-24684
                                 CIP

ISBN: 0-75961-377-X

This book is printed on acid free paper.

1stBooks – rev. 04/11/01

About the Book

Common Bonds: Reflections of a Cancer Doctor offers cancer patients and their families, friends, and health care workers a unique opportunity to examine the mind and conscience of a busy cancer practitioner. It offers a compassionate and inspiring look at a major, overlooked aspect of the battle against cancer. As such, this book opens the door to better communication and understanding and creates greater patient awareness of the physician's often hidden emotional involvement.

Taken from the rigors of Dr. Berger's daily life, including actual cases, *Common Bonds* offers a moving, illuminating narrative of wonderful, wild, and woeful accounts of doctor-patient relationships. You'll feel compelled to keep reading, even long stretches, to determine the fate of the various patients introduced.

The stories Dr. Berger provides are about people to whom other cancer patients can easily relate. The interactions he describes are certain to help readers more readily fathom their own behavior with their doctors, and thus help foster a more healing doctor-patient-family relationship.

About the Author

E. Roy Berger is a medical oncologist, researcher, and lecturer around the country. He is a partner in a highly regarded private practice, North Shore Hematology-Oncology Associates, which has offices in both East Setauket and Smithtown, New York.

Having received his medical oncology training at the Memorial-Sloan Kettering Cancer Center, he has served as a consultant there as well as having staff positions at Mather and St. Charles Hospitals in Port Jefferson and St. Catherine of Siena Hospital in Smithtown.

He serves as Chairman of PAACT – Prostate Cancer Oncology Group, and is a member of the Prostate Cancer Education Council. He has participated in numerous research studies which have led to the Federal Drug Administration approval of drugs such as Strontium and Flutamide, both considered advances in the treatment of prostate and other cancers. His subspecialty is Prostate Cancer.

In this candid medical autobiography, he shows laymen and patients what a cancer doctor thinks and feels, how strong the common bonds between doctor and patient can be, "what it's like to be the person at the other end of the stethoscope."

DEDICATION

To all of my colleagues and teachers who have helped me to better understand and treat our patients.

To all of the nurses and staff who have taught me greater compassion and love.

To my wife, Joan, and children, Allison and Jessica, who have patiently endured my long hours and difficult times.

To my parents, whose loving guidance helped me achieve my goals.

Especially, to all of my patients, who have taught me so much about myself and the life process.

Contents

FORWARD vii

ACKNOWLEDGMENTS.. ix

INTRODUCTION xi

CHAPTER 1~ Time Bombs 1

CHAPTER 2~ Seize The Day 23

CHAPTER 3~ The Courage To Change... 37

CHAPTER 4~ The Serenity To Accept.. 65

CHAPTER 5~ Taking Heart... 83

CHAPTER 6~ Rounds.. 101

CHAPTER 7~ Parallels 119

CHAPTER 8~ Making A Difference.. 133

CHAPTER 9~ The Spirit Never Dies.. 145

CHAPTER 10~ And Tomorrow's Miracles? 163

FORWARD

Perhaps it is a special combination of professional training, knowledge, skill, and ego which compels a person to be an arbitrator between life and death on a daily basis.

In *Common Bonds*, we meet one such individual, Dr. E. Roy Berger, who has selected medical oncology as his life's work in the New York Metropolitan area. In a candid, touching, and sometimes sad memoir, Dr. Berger recounts episodes in his practice from the unmasked viewpoint of a physician. His insight enables the reader to understand him as a deeply concerned practitioner, even as a husband and father, who has chosen a profession that often times has only small successes.

Common Bonds movingly describes life and death situations. It shows that there is a delicate balance in the common bonds of doctor-patient relationships which, although too often full of heartbreak, frustrations and failed dreams, can be marked by wonderful victories. Most importantly, the reader will see that the relationship between doctor and the patient is a creative process sustained by optimism, honesty, and hope.

Of particular interest to me is how Dr. Berger became one of the first few oncologists in the United States to have a subspecialty in prostate malignancy. Several years ago, he visited Dr. Fernand Labrie at Laval University in Quebec to witness firsthand a new development in treating prostate cancer. Called combination hormonal therapy (using a new drug called flutamide) it lengthened or in some cases possibly saved the lives of thousands of men with this dreadful disease. As a result of Dr. Berger's trip to Laval University, he participated in clinical trials that eventuated in the United States Food and Drug Administration finally approving flutamide as a prescription drug. Subsequently, combination hormonal therapy using flutamide can be generally prescribed by doctors in the United States.

Common Bonds depicts many true stories of extraordinary people, both doctors and patients, who together battled a common enemy — cancer.

James Lewis Jr., Ph.D.

ACKNOWLEDGMENTS

Over the fifteen years that it has taken to complete this work, there have been many, many people who have helped me in my pursuit. Of course first and foremost are the myriad number of patients with whom I have had the privilege of treating. Although there has rarely been a significant interaction in which I have not benefited, I could only single out a very few of them. To all of my patients and their families, I owe much.

The support that I have received from my caring and efficient staff and more important the support my patients receive on an everyday basis has been a most appreciated but too often presumed service. There are too many people to thank individually, but suffice it to say their contributions to patient care and, therefore, to this writing is incalculable.

The professional relationships that I formed have been very influential in helping me produce this work. There have been numerous nurses, physicians and other medical personnel who have allowed me to do a better job as a physician. Most importantly, however, are my five partners who have been at times constructively critical, but mostly supportive. Their influence has helped me immeasurably in my pursuit of attaining competency in my field and in writing this book.

Special thanks to my writer Linda Mittiga who has always strived to make this project excel. I hope we have succeeded. My publisher, 1st Books Library, arrived at my doorstep by divine intervention, and although they have my utmost gratitude, they probably don't need it.

Without the support of my family it may have taken much longer to accomplish this formidable task.

INTRODUCTION

The physician-patient relationship is a special one with many implications both obvious and subtle. Seldom is it more important to the well-being of the patient than in the area of Oncology. The emotional zeniths and nadirs — as usually ensue after a diagnosis of cancer — foster an especially close relationship between cancer doctor and cancer patient. I know this from experience.

The nuances of this physician-patient relationship differ depending upon the personalities involved. There are, however, common threads that weave through the fabric of these interactions.

In the pages that follow, I have tried to bare my own feelings, reactions, and everyday experiences as a medical oncologist. In so doing, the anatomy of the relationship between me and my patients and their families can be examined. (It is my contention that patients and families can foster an improved therapeutic relationship with someone whom they understand.) To date, most books written about the cancer experience have come from the patients or their family members.

I do not think of myself as an exceptional oncologist, rather as representative of thousands of cancer caregivers in this country. The real,* therapeutic relationships, depicted in this book, offer both givers and receivers of oncologic care a chance to scrutinize the emotions, motives, and interactions that are common to all physicians and patients alike.

The patient stories about which I have written demonstrate a broad range, as diverse as the diagnoses, prognoses, ages, genders, and personalities as any oncologist might see in practice. Most importantly, people who are touched by cancer in their lives or the lives of their loved ones will come to a new understanding and awareness of how essential it is to build a mutual alliance with their healthcare givers.

It is my deepest hope that we can all learn from these victories and failures in order to attain a better dialogue.

E. Roy Berger, M.D.

*While patient names have been changed, their stories are accurate. The story of Lynn Moline in Chapter 3 is the medical case history of Linda A. Mittiga, writer.

1~

Time Bombs

When Carl heard my words, "malignant melanoma," I could see his strong face blanch with mixed emotions of disbelief, horror, and even anger. He was fixed to his seat in front of my desk, shackled by fear.

Seated in my chair across the desk from this new patient, I wondered whether I should continue speaking. My explanations about his cancer, treatment options, and expected outcome were now meaningless to him. I could see he wasn't listening anymore. The focus of his stare had turned inward. He was wrestling with the as yet unspoken question: Am I going to die?

While he was attempting to absorb this news, my mind wrestled with its own demons. This battle against cancer was endless. Fighting this sometimes unbeatable foe was a noble but frequently futile task. Why did I pick this specialty? It was like tilting at windmills.

I mentally shook the doubts away. Caring for and treating patients was exciting, especially when current medical technology gave us a fighting chance. I could give my patients the best available treatment, and if not cure them, at least offer most of them prolonged lives. Failures were frequent in this field, but cures weren't always elusive. Over the years, there had been some wonderful victories...

Carl's reaction was typical—it was hard to learn of one's own mortality—and understandably, he was scared.

Deep down, I was scared for him too. Of all the cancers I have treated, malignant melanoma was the one I feared the most. When my mother was diagnosed with it years ago, the heredity factor put me and my children at higher than ordinary risk. Fortunately for my mother, surgery of the primary lesion and removal of her lymph nodes had cured her. Relieved and grateful as I was to have her beat the odds with an extraordinary recovery, I have never been able to stop worrying about it. My private fear of malignant melanoma has made me routinely inspect every unusual mole or skin blemish in my family.

It would have been unprofessional to unveil my personal fears by telling Carl all this. Instead, I called gently to him across the desk, a distance which might as well have been miles, and said, "Carl, listen. You feel like somebody has just dropped a ton of bricks on you. Right now, it's not as bad as all that. Let's see where you stand with this." I adjusted the glasses on my nose so that I could read his chart that lay on my desk.

He had a right to know what he would be facing in the weeks, months, years ahead, and what we, as a team, were going to do about it.

After explaining the particulars of this malignant form of skin cancer, I added, "It says here that your surgery removed all detectable cells that were located in your scalp and that there is no further evidence of diseased tissue remaining," I looked up at him reassuringly, "That's certainly something in your favor." Briefly, I hesitated before I continued. I didn't want to choke noticeably on the next words. "But, the biopsy suggests that the tissue was a secondary site."

Touching the scar on his scalp, Carl nodded in response. He was listening closely now. "What do you mean, a secondary site?"

Hiding behind a composed face, I silently reread the biopsy report to be certain before I spoke again. The truth was awful. The cells in the tumor that had formed a lump in his scalp were metastatic. This meant that somewhere else there was an undetected primary lesion. His cancer was a time bomb that could explode at any time. Who on earth was ready to hear this?

"It means that the source of the tumor has not been found," I continued cautiously, "and there is also a possibility that there'll be other secondary tumors, maybe months from now, maybe years. It's a wait-and-see situation. That's why we're going to keep a close eye on you."

Control was the key here, not cure, I thought to myself. At one time, I would have been very angry that I couldn't cure him, but after years of emotional anguish over this, I've had to reconcile myself to the limitations of the present technology. Sometimes that technology has given me tremendous power in fighting the disease, sometimes, just a little. In Carl's case, I was nearly powerless.

"I can't believe this is really happening to me..." he said almost inaudibly behind a nervous whisper. Then he erupted, "I don't believe it!" smacking his fist on the arm of the chair. "Doc, I feel so damn healthy!"

For a man about to turn forty, Carl Drake appeared much more youthful than his age. Sandy-haired, fair-skinned, his chiseled face and muscular figure matched those of Olympic athletes. His years of outdoor recreation as an amateur but avid sportsman, had wrought him a very fit body. But the excessive exposure to the sun for all those years had probably just caught up with him.

"Do I need chemotherapy?" Carl asked gruffly before I could respond to his outburst. The muscles of his jaw tightened.

2

He had to be told the truth. Chemotherapy was almost ineffective against his cancer. "It's not necessary." I answered turning the negative statement into a positive one and hoping he would find this fact encouraging. Instead his expression had turned to stone. He understood all too well.

I could see fear had paralyzed him, so I tried to redirect his focus. "Carl, you have to be positive. Your tumor does not appear to be grossly systemic at this time. That means that, except for regular examinations and tests, you will not be needing any therapy or further surgery right now. You can continue to work and participate in your customary sports activities. You see, you're fortunate, it could be worse."

Who was I reassuring here? I asked myself. Knowing how, if luck turned against me, I could be in Carl's shoes someday, the last thing I'd want to hear is some doctor droning on about how I should be grateful. I couldn't avoid it, I felt helpless.

Leaning back in my chair, I tried to shake the grim feeling of despair that gripped my gut. This would be a bargaining for time. There had been no studies to my knowledge that definitively showed chemotherapy or immunotherapy would prolong the life of a patient with malignant melanoma in this situation. However, as usual, before I ruled anything out, I would check every authority, institution, and medical journal at my disposal for innovations in treatment. Maybe there was something else to do for him. Maybe.

"So what do I do now, Doc?" Carl looked at me with a different facial expression. Now he wore a peculiar grin. It was almost a smile.

"First, I'm going to have you go to one of the examining rooms where I'll check you thoroughly for unusual blemishes, bumps, enlarged organs, or any indication of tumor development. Sometime during the week, I'll review your chest x-ray. However, as far as treatment is concerned, unless I find anything else, there's nothing to be done at this time." I responded frankly with a slight shrug. "Still, new treatments are always under investigation..." Carl's smile was becoming broader, unnerving. I stifled a shiver.

Suddenly he burst out in hideous, low laughter, a sound that raised the hairs on my spine. "Does this mean I'm dying?" he asked. Small tears squeezed out of his eyes.

This was the question I always hated to answer when the truth was yes. In Carl's case, this virile man was indeed destined for a premature death —if he were lucky, he'd have five years. Yet above all else at that moment, I hated hearing his laugh.

Death from malignant melanoma while inevitable was unpredictable. Carl would have to live out the rest of his life as if under the mythic sword of Damocles, not knowing when it would come down. As I do with most of my patients, I wanted to reach out to him at this moment of crisis to ask him how would he cope, how would he pick up with the rest of his life, knowing the time was running out so soon?

Instead, I stopped myself short. I had to shield myself from bringing his personal agony any closer. So I reminded myself that it was sufficient to help him medically and still maintain a certain distance emotionally. I would be useless to him if I let my own fears cloud my thinking. Since we didn't know each other too well yet, it would scare him too much to see his new doctor overcome by a mutual fear. Until he trusted my expertise, I couldn't let my guard down. I didn't want to jeopardize our professional relationship with familiarity.

Slowly and as calmly as I could, I answered him. "I'm sorry. The truth is that this cancer will probably kill you, but you do not appear to be in any immediate danger now. If we find and quickly treat any new growths that pop up, the better your chances. That's why I am here."

Although his laughter still echoed in my memory, Carl seemed suddenly serene. This worried me even more. Leaning forward across my desk, I asked softly, "Do you have any family—a wife, kids— you would like me to talk to?"

Speechless, he shook his head and lowered his gaze. "No." he mumbled after a brief pause, "a girlfriend, but no family around here." His large hands were interlocked in a powerful grip, his knuckles whitened. His downcast head sank even lower. He was riding an emotional roller coaster right in my office. I hardly had a chance to collect my words of consolation, when seconds later, he raised his eyes to meet mine. Something about him had changed. Staring back at me was a man who no longer seemed afraid.

I was startled. How did this happen so suddenly? I had seen patients adjust over a period of time, but not this abruptly. How was he able to find this surge of courage to face the shocking nature of his condition?

When he spoke again, his voice demonstrated a confidence that had moments before been lacking. "Doc, I know this sounds trite... It just struck me." He almost seemed embarrassed, but his gaze held mine. "It's like competition sports, life is. It's got its challenges. Just like in a game, sometimes losses are heavy. You take your hard knocks and play on. I admit, I've probably had more than my share of victories too." He chuckled to himself knowingly, "You can't beat that thrill of being the best in something. Anyway, what mattered most to me down through the years was not how often I won, but, you know the cliché, 'how I played the game?...'" He smiled shyly, and looked away.

Whatever determination made him excel in sports had just helped him best the situation now. "You've just fought half the battle right here," I replied softly, amazed by his complete transformation. He had reached inside himself and found the courage to deal with this unexpected opponent. With the attitude of a true champion, he was ready to face the greatest challenge of his life.

I could have leaped from my chair, clapped him on the back, and shouted: "That's the spirit, Carl!" at the top of my lungs. But I didn't. Instead I stood up casually from behind my desk, offered him my hand to shake, and expressed my appreciation. "I admire your fortitude, Carl. Remember, we're a team, and together we'll do our best to fight this. I can promise you that much." Behind my impassive face, I had been deeply moved by this courageous man.

When he left my office, I felt exhausted, as if I had just run in a marathon.

* * * *

My story of Carl Drake is only one of many professional cases about to be told in the pages ahead. Although, the names and diagnoses will vary, there remains one constant in each of these case histories that is of the utmost importance to me. In addition to offering my patients the highest quality medical care, I need to establish a viable doctor-patient relationship based on trust. Without that special bond which fosters mutual understanding, respect, sympathy, compassion, and even at times friendship, the practice of medicine can only be a mechanical and empty endeavor.

A beneficial doctor-patient relationship means, to paraphrase one of my patients, "the difference between peace or stress while patients are engaged in a battle for their lives. Ideally, an oncologist can become just about the patient's best friend especially since establishing genuine trust and confidence can bring comfort to the terminally ill."

For me, getting off to a right start means developing a sense of trust with patients. Unfortunately, it is not possible with every patient. While not always essential in offering them the best medical care, it is unquestionably, in the patient's own mind, the deciding factor in the quality of that care. At the same time, from my perspective, a good rapport always makes an enormous difference because mutual understanding and respect make for better communication. Better communication makes patients somehow seem easier to treat.

Yet, a sense of trust is not inherent. It is earned over a period of time. However, when the truth means a person is going to die, it is hard to provide the facts without shattering the patient's world. At times, even among the best practitioners, misunderstandings arise, perhaps as a result of a physician's apparent businesslike manner or frankness, or perhaps because the patient blocks comprehension of the unbearable news. A difficult situation becomes complicated by feelings of distrust and frustration on both sides.

Because patients deal with the bad news of their illness and mortality differently, establishing mutual trust and understanding with each patient and their families is often unpredictable. What works for one does not necessarily work for another. And if things do not go well, the patients and relatives can feel betrayed and blame the doctors.

As I relate my personal experiences as a cancer doctor, I hope that this view of one, real doctor's world will afford a rare opportunity for all patients and their families finally to understand the doctors in their own lives. If all minds and hearts can be opened to the idea of mutual understanding, the doctor-patient relationship can only enhance the power of the healing art.

* * * *

As I remained seated at my desk that day after Carl had left, I realized my face may have appeared expressionless to him. What he didn't know was that my impassivity was shielding my profound sadness and sympathy. The courage he had shown, as so many other of my patients had shown, inspired and awed me. Reminded that my fate might some day be similar, I could only hope that I would be capable of such bravery. Nevertheless, I have to let my patients believe there is some hope left to them. In Carl's case he still had good quality of life no matter what the time limitations might be. Showing my true sentiments would scare him right now. So, unless I knew how an individual would handle my unmasked emotions, I would play it safe and choke back my feelings as I had done with Carl.

No matter what my feelings, I wouldn't allow emotions to interfere with my medical ability to help my patients. Usually, I had some options to offer. With Carl Drake, however, I was torn apart by my own helplessness. How was he going to accept that he was a walking, talking time bomb, incapable of being defused? At any time, his disease could explode and kill him.

Glancing at my watch, I rose from my desk. It was already 5:30 P.M., and I still had three other patients waiting in the examining rooms.

At the nurses' station, Ruth, our head nurse, greeted me with a few questions about a patient's orders. As my right arm in the office, she had a gift for handling matters caringly and efficiently. In addition to being a great organizer, she was a warm, caring, mature person who understood life, who supported both the many, many patients in our practice and the five doctors in our group. Despite the hassles in the office, Ruth could usually coax me back to good humor with her lighthearted jokes and ribbing. Her personality and her professionalism made her invaluable to me.

Yet, the same was pretty much true about the other nurses on our staff as well. Each was a dedicated professional whose compassion and caring, and talents for handholding did much for our patients. The staff nurses were the ones who gave the chemo, talked to the patients, and truly listened to them express their concerns. They were the reliable backup, and they were good at what they did.

After Ruth and I reviewed the patient orders, Sue handed me a list of phone calls which required my prompt responses, and Marie whispered a reminder. "The patients are grumbling about how late it is."

I couldn't blame them. Nobody liked waiting. It was bad enough that these people had cancer; I disliked the thought of letting them waste precious hours. Sometimes, it just couldn't be helped. Time had to be rationed according to need. If I had to shortchange any patients timewise, usually it would be the ones who were reasonably well, the ones who were present for checkups or blood counts. On such busy days, I hoped they realized that less attention from me meant they were in good shape. Everyone would have his or her turn of my complete attention when it was necessary.

Lifting the chart of my next patient from the rack on the door and pocketing the list of callers, I walked briskly into examining room 2.

"Hello, Edna." I began, closing the door slowly. "Sorry it's getting so late. I got hung up with a new patient. Let's see how you're doing so we can speed you on your way." The door shut behind us.

* * * *

The following morning, while making my rounds at the local hospitals, I began formulating a different plan for Carl Drake's melanoma. It was becoming a personal fight.

Only slightly late for my office hours, I rushed through the side door of our Smithtown office and immediately approached the secretary. "Chris, please give a call to the University Hospital. See if they can dig up Carl Drake's original chest x-ray and tell them I'll be by late today to review it."

"Will do," she replied as she casually swiveled round to her desk. Chris was reliable, there was no need to tell her twice. She'd get it done.

I felt better. For the length of the morning, I could shelve further thought of the Drake case. There were enough patients lined up to absorb my full attention.

Lydia and her mother, Barbara, were waiting for me in examining room 3. As I entered, it was immediately obvious that Lydia was in far better spirits than she had been in weeks. She had been laughing and joking with my staff, and announced. "Oh, my favorite doctor is here!" when she saw me. Her face was pale. She had covered her total hair loss with a brightly colored kerchief, but her big eyes, rimmed by dark circles, were smiling. She had an appointment for only blood work and an exam this day, but by her happy attitude, one would have thought she had just been announced winner of a beauty pageant. In my book, she was a winner!

Over the past several weeks, Lydia had been so miserable that I was at odds about how to handle her deepening depression. The past year had been traumatic for her. She was nineteen years old and already had a cancerous ovary removed. Because it was very malignant, she was given an aggressive chemotherapy regimen after surgery which was administered once a month for five consecutive days. The chemo hit her hard. Its toxic side effects included nausea, vomiting, alopecia, and general malaise. Then, four months into her treatment, she was threatened with a setback. A second ovarian mass was discovered. We were all scared. Chemotherapy was suspended while she had an exploratory laparotomy. The mass was determined to be a simple cyst, and we all breathed a sigh of relief. While the cyst was removed, she was given a further respite from her chemotherapy protocol.

When it finally was time to resume, she dreaded therapy even more than before and became quite upset with the recommended year of treatment. Her anxiety urged me to find a less heavy-handed protocol, if possible. Checking the most current literature, I became convinced that the treatment needed to be at least nine months long, but perhaps no longer. Three extra months would make no difference in controlling the disease, but from Lydia's point of view, it would make all the difference in the world.

Three weeks ago, the Lydia I had seen wasn't smiling. Even though she had been encouraged by the shortened protocol, she was still frightened about undergoing any treatment. Since I had felt her psychological approach to the disease was instrumental in treating her successfully, I had granted her one more week to "get it all together."

A week later, amid tears and quiet sobbing, Lydia returned for her treatment. Nestling close to her mother for support when I entered the room, she went into floods of tears. It was evident that the week off had merely worsened her anxieties, not strengthened her courage. Her mental attitude had bottomed out.

Concerned about her depression, I felt more anxiety about any further delay in her therapy. It couldn't be suspended much longer without threatening her chances for survival. Shutting the door, I seated myself near them both. "Lydia," I had to remind her gently, "the sooner you begin treatment, the sooner it will end. We're talking about a matter of months from your life, not years. When this hell is finished, you can return to normal living."

This was the truth. Although, with surgery alone, only 25 percent of the untreated patients who had this disease survived five years, chemotherapy improved the figure to over 66 percent. I firmly believed Lydia had a good chance; she could live a life free of disease if only she would accept her treatment.

I glanced at Barbara. Her eyes seemed misty. While I was talking, she had reached over to her daughter and put her arm around Lydia's shoulders. Lydia continued to tremble, her head bent, her vision fixed upon her hands as they wrung a moist tissue. She had refused to look up. This hurt, but I knew she had just cause. In her mind, I was the personification of her misery. Who else was to blame for the intense suffering she endured from the chemo?

"I don't care anymore!" She broke out abruptly. The sound of her voice was clogged with the well of tears in her throat. "It doesn't matter whether I live or die." She looked up at me with swollen red eyes. "I'm always going to have to worry about cancer!" Quickly she lowered her gaze again and choked down another sob. Shreds of the saturated tissue rose to her eyes.

I plucked some dry tissues from the box near me and handed them over to her. Keeping my concern from making my voice tremble, I had to reason away her profound fear.

"I know it doesn't seem fair. You're a young girl with your whole life ahead of you." She blew her nose loudly. "While other kids your age are out having a great life, you're stuck here, sick and miserable with a life-threatening disease. You have a right to feel bad, but self-pity won't cure you." Her sniffling sounds persisted, forcing me to talk louder.

"You have to trust me! I don't want to force you to take the chemo, except, it's your best hope. I wouldn't let you suffer like this if there were any other way. Your mom, your dad, and I have only been wanting to help you all along, not hurt you." I couldn't tell if my words were reaching her. "With all that you've gone through already: the surgery, the laparotomy, the chemo. All this can bring about your complete cure. You must care what happens to you! We care whether you live or die! That's why we're all fighting so hard with you and for you."

It was heartbreaking to see her suffer so deeply from her fears. "And you won't have to worry always." I continued. "After three years, the survival curve flattens, which means, if there is no recurrence before that time, you are cured."

Lydia was silent, but I felt no victory in this. My aim had been to achieve some reconciliation with her by earning her trust. Apparently instead, she had become suppressed, without any fight left and without accepting any of my assurances.

My voice was softer when I asked, "Isn't this what you really want?" Lydia was still hunched over her hands, but her shoulders, under Barbara's extended arm, were no longer shaking. I waited for her to give me an answer, some sign that she understood. She nodded her head without looking up. I knew this would be the only concession I was going to get from her. It would have to do.

"Then, we have to start NOW to help you. We shouldn't wait any longer. Let's get it over with!" I was trying to be cheerful. "One month down, then only eight more to go, okay?"

She gave no response.

I looked soulfully at Barbara. Her face expressed a mixture of love and anguish for her daughter. It was sobering. If only Lydia would realize she was not alone, that her family had such tremendous love for her. With their support, she could find the strength to continue.

I stood and turned to open the door, wiping my eyes before I left. At the nurses' station, I gave the order to begin. Without further protest, Lydia resumed her treatment that day.

Two days later, at the midpoint of her five-day regimen, Lydia's fortitude returned. She looked haggard—she had been vomiting for forty-eight hours. Yet, her determination to weather through it had made her hostile.

"Lydia," I greeted her, "How are you doing today?"

"What a stupid question! How do you think I'm feeling?" she snapped at me. Surprised by her caustic remark, I let it and a few other cutting remarks she made during the exam pass without comment. I realized this was her way of adjusting to treatment. I had gotten worse verbal abuse from other patients.

After I completed Lydia's examination, Barbara took me aside and admitted Lydia's anger was directed at me. "Don't worry." Barbara had whispered. "Once this hostility passes, she's going to feel closer to you." I knew Barbara was probably right.

Now, a week later, Lydia was affectionate. She was giggling with her mother as I lifted her chart from the door rack. Before I had a chance to swing the door shut behind me, she was up from her chair and, somewhat flirtatiously, curled her arm around me in a hug.

"I really do like you again!" she squealed in a girlishly loud voice. She seemed ready to kiss me.

Surprised by her extreme reversal of sentiment, I was momentarily embarrassed. Normally I become uncomfortable at such an emotional display from a patient, but in Lydia's case, I was relieved she was no longer harboring any ill will against me because of her treatment. After squeezing my hand, she gleefully pulled away and returned to her mother's side.

"Well! This is some change!" I remarked happily. "I am very pleased to see you in such good spirits again!" Since my own feelings for her were sincere, I was warmed by her appreciation. I knew I would dread a relapse, not only for her and her family's sakes, but because I truly liked her. Still, the confidence I had in both her protocol and in my own gut feelings that she would pull through made me most positive about her future.

"Even with a blood count this low, you're feeling good?" I laughed. It was so rewarding to see such a turnaround. Lydia's visit would be the highlight of my day. It was a good way to start the morning.

Later, at the nurses' station, while I was jotting down a few notes in Lydia's chart, I had some second thoughts. The sense of satisfaction I felt over her projected cure momentarily gave way to some nagging questions. If she should survive, did I help her or did my insistence on chemo merely make her sick?

Most oncologists feel these doubts. Depending on the type and severity of a patient's cancer, it is possible chemotherapy is strictly overkill. A patient's recovery may be affected by some other means, such as with surgery or radiation alone, but, because it is never a certainty whether a particular patient would have remitted without chemotherapy, the conservative authorities recommend combination therapies.

There have been helpful reports on randomized treatment (chosen statistically by computer) studies and controlled studies in which there were treated versus untreated patients. But even these studies could not give 100 percent guarantees on their findings. For any given disease, while one group of chemotherapy users would show a definite advantage over the other group of nonusers, within each group there are those dreaded, unexplained exceptions. Either, for more unfortunate patients, chemo fails to destroy their disease, or for less unfortunate patients, chemo is an unnecessary and added risk in obtaining cure. Since there could be no way to determine to which category of exception an individual may belong, most oncologists push for chemotherapy. It has been the favored alternative.

"Doctor Berger." The nurse handed me a chart. "Mrs. Lynstrom is in room 4." She was leading another patient into room 2.

* * * *

Later that morning, between appointments, I called the melanoma specialist at Memorial, Doctor George Winston. Hearing myself outline the details of Carl Drake's case, I realized how personal my fear about his melanoma was becoming.

George listened, questioned me about a few points, and then offered his advice. "Outside of an experimental immunology protocol, " he began without hesitation, "there is very little more that you can offer this man than you have already done." This was reassuring news in one sense, devastating in another. "I agree with you; there is no point in starting him on any treatment. You'll just have to wait for recurrence and treat him then for the metastatic disease." He paused a moment. "I know it's disappointing, Roy. I suggest you contact our immunotherapy group to see what they can recommend." He gave me some names, and after an exchange of thank-yous, we hung up.

At the end of the day, I made the extra trip as I had planned back to the hospital. Earlier, when Chris had called them, radiology had pulled the chest x-ray, done at the time of Drake's initial surgery, and told her it was negative. I had already known that. This information was in our medical reports at the office. However, I had wanted to be absolutely sure that there were no metastatic lesions in the lungs which may have been inadvertently overlooked. Ordinarily, I wouldn't need to double check the opinion of the radiologist, but, in this case, I couldn't rest without seeing it for myself personally.

Hunting up the x-ray was a problem. The x-ray had been mislaid. Since it was the close of the regular business day, no one was available to help me locate it. Frustrated by this setback, I went to pathology instead. The chief of pathology looked at the lesion with me. While I was there, he called the emeritus professor of pathology at the University for another opinion. Our discussions opened up a new possibility: If the lesion had begun in an appendage below the scalp line, a further and wider excision in the area might be curative. Although, I didn't believe that was the case, it was certainly worth considering. I would have to check it out. Thanking the pathologist for his time, I left. It helped to have the perspectives of other specialists.

Mulling over the events of the day and in particular Carl's case, I drove home. It was late. My wife and daughters had probably finished dinner. If I were lucky, I could see the girls through their bath time and tuck them into bed. Home tonight also meant phone calls—I was on call. Disruptions were inevitable, and as long as Joan could be patient with a fragmented evening, so would I.

When Carl phoned my office a week later, I had only a few more theories to provide. There was nothing really new or satisfying to help him.

"I'm grasping at straws," his voice, though steady, seemed desperate. "Should I seek another opinion? There's a surgeon in the city who specializes in melanoma."

His questioning didn't upset me. I was confident that I was doing as much as anyone could. "Honestly, Carl, I don't think you need more surgery just yet. I'd rather look into having the lesion sectioned again. There is also a tumor conference scheduled on it. When I get further news, I'll let you know what your choices are. If then, you feel the need to go to the melanoma surgeon, or if the conference determines this is necessary, we can arrange the surgery as you wish."

"I'm sorry, Doc," he said. "I don't mean to be any trouble. I just can't help wondering what I should do."

"You don't have to apologize to me." I answered. "You have a right to explore every opportunity to ensure your chances. It's all very normal the way you're feeling. In fact, personally, I think you're handling this whole thing very well. Just remember, you can trust me."

"Thanks!" he paused for a moment. Suddenly the floodgates seemed to open for him, and he began to speak rapidly. "Usually, I'm just an easygoing guy, almost to a fault. Being in limbo like this about my whole future has really put me on edge. I never appreciated what it meant to be young and healthy. Who ever thinks their life is going to end? I mean, maybe when you're old or really sick for a while you start to consider it. But, I haven't ever been this sick. Ironically, I don't even feel like I'm dying now either..." he was overcome by so many contradictions.

"Well, you're not dead yet, and as long as you feel good, you should take advantage of it." I was thinking what my own alternatives would be were I in such a situation and heard myself say aloud, "You know that TV program in the 60's, with Ben Gazzara—what was it called? It's not a bad idea to do things you've always wanted to do, to 'seize the day.'" I was saying things without hesitation, things I wouldn't ordinarily have said. I was completely empathizing with his fears. Somewhere in the phone conversation, I had dispensed with the doctor-patient roles, and was consoling him as if he were myself.

"I know, I know. That's what I've been feeling! There's a whole world out there. I haven't done half the things I want to do. Sometimes, I feel like shouting and maybe, maybe I'll wake up from this crazy nightmare," he sounded choked up.

"Is your girlfriend supportive?" I asked.

"Sarah? Oh yeah, she's okay. Sometimes she seems shocked, and a few times I caught her crying. I guess she doesn't want to show me how scared she is. But sometimes she's very strong. We're managing."

Automatically, I cast a glimpse at my watch and realized I would be unable to continue the conversation much longer without falling way behind schedule. "What I would suggest," I began switching back to the doctor role somewhat abruptly, "there are some therapy sessions sponsored by our office which may be helpful to you. You can discuss how you feel for as long as you need to. If not in a group, at least see the therapist personally"

"Well," he hesitated. I wasn't sure if my advice or my persona switch had given him reason to pause. "I don't like shrinks." He was retreating. Had I scared him off?

"She's not a shrink, she's a social worker and therapist." I explained. "She's not going to analyze your psyche; she's there to listen and to help you cope."

"My best therapy is to go off somewhere and not think," he admitted. His voice was growing more distant in the receiver.

I thought to myself, denial would be effective for a while but not forever. "It's there if you need it, Carl. You don't have to go. But you can't hide from this too long. At least for now, stick around until we have your problem labeled. After that you can go out and enjoy yourself as long as you feel all right and check in regularly. As for the conference, I'll let you know when I hear anything further. Give me a call in two weeks."

"Thanks again, Doc. I appreciate what you're doing." I heard the receiver click. A feeling of helplessness washed over me.

It was over a week later, when his case was presented to the tumor conference. As I suspected, experimental immunotherapy was ruled out. There was no experimental protocol currently available that would make his chances any better. Pathology, however, had started the ball rolling on a new course. Although most were of the opinion that the lesion was definitely metastatic, it was agreed that slicing multiple sections through the lesion would be informative.

After the conference, I ordered the sectioning for every few millimeters in the lesion to be sure that the epithelial surface was not broken. My strong convictions about the original diagnosis made me feel certain of the findings, but

I hoped I wasn't closing my mind to any other possibilities. If I were wrong, if the lesion was indeed a primary not a secondary deposit, then Drake's prognosis would be much better. Although I would feel bad and inadequate to have made such a incorrect evaluation based on the facts at hand, I would feel happier for Carl's sake as well as convinced that this out of the ordinary follow-up was worth the expense of time and money.

Since it would take some time for pathology to perform the section, I again had to advise Carl to give me another week for the results. These constant postponements were tedious enough for me; I could only imagine how nerve-wracking they were for Carl.

It was early November when Carl called the office again. The timing was particularly bad—a patient of mine had just died in the hospital—and Carl's pathology results were still pending. After a brief explanation, I asked him to call me back later that day for the information. By that time, most matters would be settled and his lab results would be in.

It was very late, 6:00 P.M., and he had not called. Maybe he sensed the outcome. The results left no doubts. There was no epidermal involvement. The lesion was secondary. As much as I could rest easily that everything was done appropriately for him, Carl's chances were terrible, about 5 to 10 percent survival at five years. No wonder he was afraid to get the news.

I liked Carl. In the short time I had come to know him, he struck me as an amiable fellow. It was going to be hard telling him the full truth. I rang his home several times before I finally left the office. No answer. I would have to call him later. Mentally reviewing the day's events during my drive home, I became depressed. It was not altogether a typical day. There seemed to be more misfortunes than good fortunes for my patients, and some of them were people I cared very much about.

Even though oncologists experience the ups and downs of life and death struggles on a daily basis, some days it is harder to cope than others: lung cancer, inflammatory breast cancer, lymphoma, liver cancer, colon cancer, the types and varieties are endless..., the diagnosis repetitive for many of them: aggressive, progressive, terminal... That fear in the patients' eyes when these words are uttered is heartrending. The harder it is to tell them, the higher the physicians' defenses rise. In order to survive emotionally, doctors have to learn to switch their genuine involvement on and off. Flip the switch "off" and a guard shield rises to protect the heart. Caring momentarily becomes merely an acting role. The problem arises when a doctor gets stuck in either switch. Emotion can distort the intellect, but an absence of emotion destroys compassion. This is a common dilemma among practicing physicians.

A number of times this day I heard myself make that switch in my tone of voice; I heard myself deliver terrible news in a matter-of-fact manner. When my patient's misery was beyond language, beyond words of kindness or compassion, all I could do was reach out to hold a hand or pat a shoulder to demonstrate in some way a gesture of comfort, but I could not give any more. My whole self was missing. It was too much for too many in one day.

Sometimes, this happens out of fear. Because I know from experience what is in store for them, I fear for them more than they fear for themselves. The disorderly growth that is cancer, and the inability to control that growth once it shows its resistance to treatments, leaves us, both patients and doctors, helpless without options.

When I walked in the door, Joan saw by my face that I had brought my troubles home with me. After twelve years of marriage, she knew me well. The best medicine to bring my humor around lay in her secret formula, a refreshing concoction of Allison and Jessica. My daughters were bathed and waiting for me to say good night before going to bed. Their chatter and laughter made me momentarily forget the day's woes. Begging for a chance to take off my jacket and wash my hands, I guided the three of us to the bathroom first before we hurried to the bedroom.

"Read us a story! Please?" My youngest grabbed my hand and led me to her bed. After a whole day of handholding to comfort the sick and dying, no hand compared with hers. Her delicate little grip filled me with such peace. Contentedly, I played, laughed, and read some stories before I tucked them into bed. After exchanging good night kisses, I emerged from their dark room and headed for the bright kitchen where Joan was heating my dinner.

Although my temperament was improved by my daughters' attention, I still felt out of sorts. I remembered that I had to call Carl. This night, I was finding it hard to handle disappointments. I did not want to have his tragedy invade my home. This was not a night on call, and in addition, I had planned to plow through the piles of medical journals beside my night table—burning the midnight oil— to reduce the vast store of current medical knowledge that was slipping by. This was an overwhelming task in itself.

And there were those other upsets from the day. It was all coming back to me.

"Are you feeling better?" Joan asked me while I blinked from the blinding kitchen lights.

"Better." I mumbled sullenly, seating myself at the kitchen table.

She removed my heated dinner from the microwave and slid it before me on the table. Automatically, I started eating without noticing the contents in my plate. "Anything wrong in particular?" She questioned encouragingly. A nurse by profession, as well as a terrific wife and wonderful mother, Joan possessed that great capacity for caring deeply and knowing how best to assuage the hurts of her charges... including me. I have often thanked God for her gift of understanding, especially on a night such as this one. My mood was growing ugly. The reason wasn't too far from the surface. Joan knew just how hard to dig.

"Well, there are a few things." I forked angrily at the food. "I have to call a new patient tonight with some bad news. I'll do that after dinner. Also, today, one of my patients, Greg Kenny— you know, I've told you about him, " She was nodding her affirmative, "I have been treating him years now for low-grade lymphoma. Well, he died today." I paused when I heard Joan moan softly. "It was coming. I'm not surprised, but I've really grown fond of him and his wife and kids. I just visited him a few days ago in the hospital too."

Joan's sympathetic remark escaped me. I was already recalling the unhappy circumstances that led to this day.

A number of years ago, when Greg Kenny first came to our group for a bleeding abnormality, one of my associates did the workup and treated him for what he later determined was a mild problem. Greg Kenny was a vital young man in his early thirties. His condition at that time gave our group only marginal cause for concern. As anticipated, he recovered without incident.

Misfortune did not wait too long to strike again. A few years later, he went to a throat surgeon for a biopsy of a swollen gland in his neck. The subsequent diagnosis was a different and worse condition—low-grade lymphoma. This kind of lymphoma was hardly a dire emergency. It could be controlled by treatments, nor was it immediately life threatening, but inevitably over a period of five to ten years, the disease would persist. After numerous recurrences, it would eventually kill. The duration of effective treatment was variable. Some patients with this disease lived for many years unhampered by the cancer. Others do not fare so well.

Once again, Greg Kenny was scheduled to return to our office, but the conditions and course of action had changed. In expectation of his first visit, we included his case in our Friday night conference where my associate reminded me of his health history. In addition to this background, I reviewed the most current journals on the subject, hoping my knowledge could be better enhanced with insights on the newest research. There was no new glimmer of hope.

As I had already known, researchers were divided in their findings. A low-grade lymphoma of this type could be treated in one of two ways. Most medical papers support the chemotherapy regimens that would keep the lymphoma in control or remission with doses sufficient to shrink the chronic tumors, but with the knowledge that the survival curve is one of always recurring, the general consensus being that there is no cure and relapse is unavoidable. At best, steady treatment with each relapse would curb it for a total grace period of under ten years. To the contrary, the minority opinion had strong arguments for an aggressive treatment. Since such protocols cure the more aggressive lymphomas, a similar regimen could increase the chances for actual cures in patients with low-grade lymphomas.

Wavering between the two opinions, I had felt some anxiety about making my decision without the patient's feedback. Not knowing Greg beforehand, I was also anxious that if he turned out to be a skittish or uncooperative patient, he would be more of a hindrance than a help and my decision would be that much harder.

When he was called to my office that first time to hear the strategy of treatment, Greg's reaction was most remarkable. A man of incredible determination and strength, he did not flinch with the news. Asking the appropriate questions, he seemed ready to confront his problem head-on.

"How serious is this?" He phrased the question as if it were a statement, while his voice was steady and controlled. Greg's pleasant face seemed composed, his long arms and legs calmly remained in perfect coordination with the rest of his relaxed body, nothing appeared frantic or nervous as he sat before me. Only a lock of hair, that hung in a chestnut curl over his forehead, seemed untamed.

"These results are not great," I began, surprised by his calm acceptance, "but compared to other kinds of lymphomas that are aggressive, this low-grade lymphoma has a survival time which is much longer, except there is no cure. Your condition is manageable however, and remission is possible. For many years, your quality of life can be good." Later, I was to discover how much he had to live for: a lovely wife, two terrific children, a great job as president of his own advertising company. He had everything he needed to make his life happy— everything except his health.

His attitude seemed good. "What do we do about treatment?" His gaze was commanding and his hazel eyes were fixed on me.

While explaining to him the conflicting opinions about recommended treatments, I voiced my own feelings of trying for cure with an aggressive protocol. "I have to be honest when I say that this is not the usual plan. More than likely, should you decide to ask around, you will find that most doctors would probably be comfortable with recommending the less aggressive approach."

He listened and nodded. "Your reasoning makes sense to me." His hands rested quietly on the arms of the chair. "And although the risks are there as you've explained, I have to agree with you. It's worth taking the chance." His steady gaze dropped slightly and then he looked up again. "Do I have time to consider my options?"

"We have some time to play with, but I wouldn't recommend too much delay. It only makes it harder to treat." We discussed his alternatives as if we were outlining an advertising campaign, not his life, but my fears that he was denying the true nature of his condition quickly were dispelled with his general line of questioning. He comprehended completely and was not daunted. Instead, he seemed realistic about his chances and eager to do his best, regardless. I was awed by his courage.

When he rose to leave that day, I stretched out my hand in admiration. "Thank you for being so brave." I shook his hand firmly, admitting candidly. "I appreciate how well you are taking all this. It will make fighting your lymphoma much easier if we're working together."

He smiled confidently. "Well, there's every reason to hope."

But future trials would test his mettle.

At first, Greg told only his wife about the prognosis, and even then, it was a few days after his initial visit. Perhaps it was out of a noble sense of Spartan manliness or else from a need to keep his personal concerns private, that he tried not to unload his burden on anyone else's shoulders. Indeed, as president of his own business, he was concerned with appearances especially as they would affect his clients and personnel. Whatever his motivation, he was capable of maintaining this courageous front for as long as his illness could remain hidden.

Chemotherapy does not keep secrets well. Its side effects have an insidious way of making their presence known. Hair by hair, alopecia became obvious; nausea and vomiting were difficult to hide. Despite his willful refusal to succumb to their effects, they were stronger than he. His growing incapacity became all too apparent, making him generally incompetent to carry on. Business could not continue as usual. Reluctantly, he had to step aside and concede to the overpowering strength of his treatments.

The side effects grew worse, beyond his endurance. His handsome appearance altered and his courage wavered. The knowledge that it was still only a gamble made his determination falter. He began skipping appointments, not showing up for necessary treatments—and I grew angry.

Calling him at home, I let him know my great disappointment, "You're blowing it! If we're going to have any success, you have to do your part and follow the protocol exactly as we planned. Come on, you've got to play ball long enough if it's going to work."

"I feel like a damned guinea pig!" he grumbled heatedly. "Chemo is too hard. Is it worth the risk? Is it worth the sickness? I feel worse now with the treatments than I did when I was 'sick' with lymphoma."

"I know it's very hard. What's worse, I can't honestly answer your questions either. Only time will tell if all this is worth it. As I explained to you before, we are experimenting. We are not doing the ordinary protocol. But give us just a little longer before you throw it all away. I won't let you go through all this if it has no chance of working. Trust me. We need a little more time. You've got to come tomorrow."

Always a reasonable man, Greg persevered. Gradually the disease melted away with only a few suspicious nodes remaining. At this point, we had weathered through the induction phase together, but not well. Greg was too battered and weary to continue to the more intense consolidation phase, and meantime, I was convinced the aggressive approach was not going to work. Greg was nearly NED, no evidence of disease, but nearly in this case was not good enough. Those suspicious nodes were an indication that the regular course of low-grade lymphoma therapy would have to suffice.

Within a few short months after treatment ended, Greg relapsed. Controlling it was no easy matter. But we managed to give him several years of this cat-and-mouse game before therapy began to fail him completely. They were not easy years for him, either, but they gave him time to watch his children grow, time to enjoy the happier sides of life which he did to the fullest.

After four years, the disease spread rapidly throughout his body. In my experience as an oncologist, I had never seen a low-grade lymphoma take over its human host so swiftly. He had neck masses the size of softballs, in x-rays, his groin and chest areas showed innumerable tumors, and they were multiplying. My alarm at the explosive manner in which the lymphoma took over his body was hard to hide. I dreaded his office visits as much as he. We both recognized that desperate measures were our only recourse.

Futile as it would seem, I would sit down with him each time in the office and talk about his next alternative. Together, we would explore the effects of heavier treatments or the risky possibility of a bone marrow transplant. And bravely he would suffer each attempt hoping desperately for success. As time went on, greater doses were too devastating, and no donor was found for the bone marrow. I tried everything I could muster on his behalf. He got sicker and sicker while the disease grew stronger.

In the last few months of his life, fluids swelled threateningly throughout his body and the pain was unendurable. He had to be hospitalized for his own comfort. Despite his protests and my own personal feelings in knowing he would never go home again, I worked to convince him of the benefits in being hospitalized. Nowhere else could he be provided with the constant care he needed. The tragedy of separation from his family was lessened only by this necessity. Always a reasonable man, he conceded.

It was only a few days before he died that I talked with him in the hospital. While excessive body fluids drained out through the tubes in his chest, a morphine drip gave him comfort from the immeasurable pain. Weakly he smiled at me when I entered his room. It gave me the chills to see him work so hard to form a smile. The hazy gaze of his sleepy eyes told me the morphine at least was doing its job.

The question "How are you doing?" was never applicable in these situations; rather the more appropriate "Do you feel any pain?" or "Are you comfortable?" was all I could ask.

I sat beside him on the bed and patted his hand. I found my words were sticking in my throat. "Is there anything you need to make you more comfortable? What can I do for you...? " I asked quietly.

Still with a hint of command, his eyes were fixed on me. "There is." he replied softly. I leaned closer to hear him better. "You can send me home. I want to go home."

"Oh," I groaned, "I'm sorry, Greg. You know how difficult that would be for you and Karen and the kids." I was stumbling for words because it hurt so much. "I can't, I really can't let you go." He was dying, but if I sent him home, he would die in excruciating pain in the presence of his very impressionable children. His death would be even more traumatic for them. Around-the-clock nursing would be financially exorbitant because professional homecare and hospices were not yet available for ordinary patients like Greg at this time.

He nodded at my response and closed his eyes. "I knew that would be your answer." He exhaled loudly, before he continued, "but I had to ask anyway." Then he opened his eyes again and held mine in his glassy stare. "You know," he began, "if circumstances were different, we would probably have been friends."

Caught off guard by my own emotions, my eyes welled up. "You're right," I admitted sadly, looking back at him through blurred vision. "I think you're a terrific guy. I've admired your courage, and strength in the way you have handled all this." Words were not doing my feelings justice. "You're... just such a... fine person..." I was choked up and could not continue. Instead, I held his hand a little longer until I regained my composure. Then, I rose to leave. "Bye, Greg. I'll be back."

When I had turned to have one last look, he was smiling at me.

"You seem to be taking his death pretty hard right now," Joan observed after I recounted the details of Greg's passing.

"It's just harder to cope when you feel there was an opportunity missed somewhere—we could have been friends—and in one sense we developed a friendship while trying to combat his lymphoma, like army buddies. Then, just when I have lost one, I have to tell another he's going to die soon." I removed my glasses to rub the fatigue and grief from my eyes.

"You mean this new patient you have to call tonight?"

"Yes! Look at the time! I've got to try again, now." Shoving my empty dinner plate away, I hopped up, gave Joan a thankful kiss, and whisked away to my den. I realized my sadness for Greg Kenny was alleviated by the thought that he was now out of his misery. Such consolation was the minimal trade-off for such a great loss.

After swiftly dialing Carl Drake's number, I heard a woman's voice answer, "Hello?"

"May I speak with Carl, please? This is Doctor Berger."

"Oh, yes. Just a minute," she answered politely.

In seconds, I heard Carl's familiar voice in the receiver. "Hi, Doctor Berger."

As I was about to speak, I realized my pessimistic mood had shifted. Carl was not dead yet. He was still alive and capable of speaking to me this very minute on the phone. Who could explain the arbitrary fates in the lives of people? Greg had died, but Carl was still alive! I had to keep my voice from sounding overjoyed. The news I was about to deliver was bad, but I wanted to remind Carl that he was lucky to be alive just the same and that he should make the most of that variable length of time he had left. He should not despair. He was not alone in his mortality. Everyone was a time bomb of sorts. This was our common fate. This was the human condition.

"I'm sorry to be calling you so late," I began as pleasantly as possible without sounding inappropriately optimistic. "I know you had wanted to hear the results of the sectioning."

The dice had been rolled. Life and Death were gambling for another man's fate.

2~

Seize The Day

The most frequently asked question that I hear from both medical and non-medical people is "How do you do it?" I guess the thought of dealing with death on a daily basis makes most people feel uncomfortable, unprepared, ill equipped and downright scared. This fear of death spills over in any discussion about an often incurable disease like cancer and so we use euphemisms to mask our concerns.

But being an oncologist never allows me to forget my own mortality. I face it every working day and certainly, the message—cancer knows no age boundary—-becomes clear. I have seen it run its gamut from very young children to centenarians. There are no euphemisms to mask this reality.

Whereas most people take life for granted and expect that they will get up each day as healthy as the day before, my experience shows me that this is not always true. Instead, I have learned that each day is a gift, to be treated as such (this is what I had tried to tell Carl Drake that night), and that time should not be wasted on useless things. The more I deal with cancer patients, the more I feel a pressure from within to appreciate each day and to try to accomplish goals as soon as possible, to do the things that make me happy, and spend my time with the people who mean the most to me: my wife, daughters, parents, sister, and my extended family of close friends.

Oncology has helped me recognize the finiteness of life, but I didn't start in medicine with this philosophy. It developed over the years, and probably began at a turning point in my life, the summer of 1970.

* * * *

I met Joan on Fire Island, the third weekend in July, three weeks after I started my internship at Roosevelt Hospital in Manhattan. Neil, an old college friend, who rented a house in Bayberry Dunes (which has since been converted into a state park), had invited me to meet another member of the house that weekend. I had agreed to come, and as soon as I got off my shift, made the connections to catch the late afternoon ferry.

My blind date was a disappointment, however. Wearing an overly large hat which flapped in the lazy afternoon breeze, she was lounging in a beach chair when I approached the side gate. Neil spotted me standing there and beckoned me in to introduce me to Jasmine. She nodded a silent hello and then offered me a taste of her drink, a pureed fruit concoction of her own making, but I politely declined. It was a muddy chartreuse color and didn't appeal to me. A lame conversation began. Jasmine turned out to be a yoga-oriented, chanting, health-food type with an anti-medicine, self-healing belief system that was not going to mesh with my personality or style. This I discovered within the first hour. I spent a few hours trying to figure out how to get away politely.

Before I had to make a decision, however, Joan (who also had a share in the house) and her girlfriend Deb arrived on the last ferry. "Sorry we're late!" they both announced in unison, appearing at the side gate, appropriately festive with parcels and bags for the weekend. As my mismatched date walked off—-and out of my life forever—-Joan walked in. "We missed the ferry and had to wait two hours!" She laughed good-naturedly, exchanging amused glances with Deb while the house members relieved them of their packages and gear.

It was a cool July evening with an ocean mist heavy in the air. The sun had already set, but the sky remained a deep crimson. Dinner was about to be cooked on the charcoal grill, and flames from the newly lit fire were shooting up, lighting the faces of everyone greeting Joan and Deb.

Introductions were made and all Joan's friends exchanged kisses with her. I wasn't shy. "How about me?" I said and received a friendly kiss on the cheek from this very warm, friendly, and attractive woman, destined to become my wife.

After dinner, friends and acquaintances divided into groups to go down to the beach for a walk. I was tired from long hours at the hospital and preferred to continue the conversation I was having with Joan, so we decided to stay back at the house. Resting as comfortably as possible on the cramped sofa, I enjoyed listening to Joan's soothing voice. I learned she had been a nurse in charge of an obstetrics ward at Mr. Sinai Hospital and was currently working for three plastic surgeons on Park Avenue.

We talked for about an hour despite the weariness that hung over me. Evidently, I must have drifted off into a deep sleep, because the next thing I knew, Joan was nowhere in sight and it was a bright, new morning. I had slept soundly through the whole night. While I slept, Joan had put a quilt over me and a pillow under my head.

I was touched and impressed. I felt then that this woman was a kind, considerate, and warm human being. Those feelings have never changed, only grown with the passing of years.

About a half-hour after I woke, Joan emerged from one of the bathrooms, bright and clean with her long brown hair still wet and stringy from her morning shower. She was a vision of loveliness wearing a two-piece, purple bathing suit. Infatuation began at that moment: first my eyes followed her around, then my body followed her all day at the beach—until I found myself willing to follow her for the rest of my life.

We quickly fell deeply in love. Since it was difficult to juggle our free hours together and intolerable to be apart, it was not long before Joan moved in with me during my internship. New York was a romantic city for lovers and set a splendid backdrop for our newfound relationship. We totally enjoyed spending our time together no matter what we did, whether it was walking in the park, eating at restaurants, going to concerts, ballets, shows, or making love.

After we had lived together for three months, I proposed to her, and when she gladly accepted, we set our wedding day for March 21—the first day of spring. Both our families rejoiced in our happiness.

Getting through that first year as a medical intern was rough, but I knew it would have been more difficult for me if Joan had not been there. She was then (as she is still) a great source of emotional support. I also found I could rely upon her intelligence and experience as a nurse to give me insights on what patients were actually feeling when the doctors left the room. It was frequently the case that veteran nurses, such as Joan, knew much more about some aspects of patient care than young M.D.'s fresh out of medical school.

Professionally, my growth after medical school during my years as an intern, and after as a resident, was also due in large part to the bonds that I formed among the dedicated doctors and nurses who shared their experiences with me. Many of them, down through the years, have become my colleagues and friends. I realize I have been fortunate to benefit from the great diversity of their intellectual stimulation and this interaction has added dimension not only to my career but to my personal life as well.

Among the many wonderful people in the medical field whose friendships I value, there is one particular person, Bob Hickes, whose dedication greatly impressed me when I was his resident. Our friendship grew from my professional respect for him while he was my intern, and since then, we have maintained a warm and close relationship for more than twenty years now.

One afternoon, back in our training days, Bob admitted a patient who had a persistent cough for three or four weeks, was running fevers, and was obviously suffering from some kind of pneumonia. Randy Seaver, a long-haired flower child from the sixties with multicolored, stained teeth and a bushy red beard that hung down to his chest, had recently returned from traveling through the Himalayas where he and a group of friends were helping to feed poor Nepalese children. Sudden drastic weight loss of almost thirty pounds combined with the other symptoms made Randy aware that he needed medical attention.

Bob presented Randy's case along with the chest x-rays to me, his resident, and we discussed the possible diagnoses. Bob suggested tuberculosis but I was skeptical. It seemed a remote possibility since the area of infiltration, evident in the films, was very atypical for TB. Bob, however, felt TB was high on his list of considerations and was determined to prove his case to both me and our attending physician, if not by dinner time, then he vowed by the next morning. When I left for my apartment with the evening off, Bob was still in the lab.

"Roy, come over here right away!" the voice on the phone replied to my irritated hello at 11:00 P.M. that night.

"Bob?" I asked, "What's this all about?"

"I found a red snapper!" Bob yelled excitedly, "...on the acid fast stain, on Randy Seaver, just now..." He was understandably thrilled. "I want you to come over here and verify it."

"I'll be right over," I replied incredulously, and then hung up.

"What is it?" Joan asked sleepily as she watched me hurriedly gather my clothes and get dressed.

"I have to go over to the hospital," I explained, "to see if that crazy intern, Bob Hickes, has really found Mycobacterium tuberculosis, the causative organism of TB pneumonia..." I was talking quickly and just managed to finish my explanations before I raced out the door.

Shortly after, when I arrived at the lab, Bob was still bent over the microscope. He jumped up when he heard me enter, rushed me over to his microscope, and pushed me down for a look. "See it!" he said anxiously. "I have it dead center in the field."

He was right. Fixed in the fluid under the magnifying lens was a straight rod-shaped, red-stained bacterium usually identified with TB.

"I don't believe you!" I exclaimed in wonder. "How long have you been searching for that organism?"

"About two hours!" he said proudly.

"Is that the only one you found?" I asked even more impressed by his perseverance.

"Yep, what do you think?" he asked, confident about my reply.

"I think you're crazy!" I smiled broadly at him, "but it sure looks like a tubercle bacillus to me." This was indeed an accomplishment, and I had to admire him.

"Then, we can start therapy on Randy tonight," Bob said impatiently, seemingly ready to head out the door.

"Wait a minute," I stopped him. "We can't commit this guy to months of treatment based on one lousy red snapper, at least not unless the attending endorses it. Let's wait 'til we show it to Shuman tomorrow, and hear what he recommends." I reminded him, trying to curb his enthusiasm with a touch of realism.

When Bob's face fell with my remark, I softened my practical edge. Patting him on the back, I reassured him, "You seem to be right on the mark here. I know I'm impressed! I can't believe you spent so much time relentlessly looking at these slides."

"I just had a feeling this guy had TB," Bob said with determination still in his voice, "so I had to prove it and I did!" he concluded triumphantly.

"Well, I'm surprised at the findings and amazed at your endurance." I beamed happily at him. "Let's see what Doc Shuman says on rounds tomorrow. Put a sign on that scope so no one moves the slide and I'll see you in the morning." Despite my own excitement over his discovery, I couldn't shake off my fatigue from the long day. "Good night and good work, Bob!" I clapped him on the back one last time as I left the lab.

The next morning, Bob presented the case to the pulmonary attending who also had felt that TB was very low on the list of possibilities for Randy's pneumonia. Shuman needed to be convinced. At that point, Bob triumphantly brought us all down to the lab to see the acid fast bacillus under the microscope only to discover that someone had ignored his warning sign. The slide was gone and Bob's evidence was lost.

Frustrated and greatly disappointed, Bob felt defeated.

"It's true, though. Bob called me last night to see it myself," I offered, dismayed by the unfortunate turn of events and feeling extremely sorry for Bob. "I did see the red-stained oblong bacteria he found."

Yet, without firm documentation, Shuman felt he couldn't condone the long and arduous therapy for TB. It wasn't a complete loss, however, because we all agreed that if no other diagnosis was forthcoming during Randy's hospitalization, we would start him on antituberculosis therapy. It was better to take this risk and administer treatment than have Randy go untreated for the four to six weeks it would take for the cultures to confirm or deny Bob's claim. After a week, with no other diagnoses to explain his condition, Randy was discharged on two antibiotics for TB. He was quite relieved to begin his road to recovery.

One day, a month later, Bob came running up to me and shoved a bacteriology report in front of my nose. "Read this!" he said proudly.

"Mycobacterium tuberculosis on Randy Seaver. I think that's great!" I applauded him. "Congratulations. Let's give Shuman the news."

Over at Shuman's office, Bob showed him the Bac-T report. Amazed and happy at the results, Shuman and I expressed our admiration for Bob's faith in his medical knowledge and his determination to prove it.

My admiration for Bob continued to grow from this point on and began a steadfast personal friendship that has endured.

My training years with Bob and my other colleagues went quickly as I progressed through residency, and soon enough I was at the crossroads of a career decision. Joan and I considered it carefully since our future life depended on the decisions made then. My two choices were to become either a general internist or a surgeon. While surgeon's hours and extra years of training didn't appeal to me, (and Joan agreed I was not enough of an early riser to hack that discipline), I felt that I was too much of an obsessive-compulsive to become a general internist. This was mostly because I didn't feel that I could comfortably keep up with so broad a field. Instead, a subspecialty as an internist would appease my desire to try to know everything and yet not overwhelm me with an impossible volume of knowledge.

Choosing a subspecialty from all the areas in medicine with which I had some familiarity and interest was not easy, but hematology seemed to be the most intellectually stimulating. One of my preceptors once said, "The blood flows through every organ." He meant that blood dyscrasias (abnormalities) had important effects on the body. One needed a full understanding of human physiology to become a good hematologist. This appealed to me, but, at that time, hematology wasn't broad enough a field to allow me to go into a private practice, something which was always my goal.

However, the field could be broadened. I knew that most hematologists at that time were also treating patients with neoplastic diseases—-in other words, cancer. The possibility of expanding into this second, related subspecialty of medical oncology was attractive because it would then allow me to go into a private practice as I had wanted. Yet, I was somewhat concerned about what it would be like to treat cancer patients. I knew that most of them died. I was not sure if I had what it took to deal with people day after day with, by and large, hopeless illnesses.

Although this area would require extra years of training, I began to feel it would be stimulating to be on the threshold of such an important and worthy venture. Already, there were so many important advances being made in the treatments of leukemia, lymphoma, and breast cancer that I finally decided to investigate further. This career decision, however, meant moving to the east side of Manhattan to live in staff housing at Memorial Sloan-Kettering. There, I could be taking an elective in my first year as a medical resident. I felt that being at an institution solely dedicated to the treatment of these disorders would enable me to decide if this was the right career choice for me.

I spent three months on the acute leukemia service at MSKCC with Doctor Timothy Gi. Tim was then and is still a dedicated leukemologist who taught me a great deal. It was during this time that I saw the tremendous impact a physician could have on another fellow human being's life. Patients would be referred to Memorial with newly diagnosed leukemia and I was moved by the real-life drama being played before my eyes. What stirred me was the simultaneous feelings of tension and excitement as the physicians raced against death when treating this disease—-each patient's life was at stake. What touched me deeply was the anguish and frustrations which the patients and their families experienced in trying to deal with a frightening disease that had so suddenly invaded their lives.

At best, what we all had hoped to achieve was cure, at worst, prolongation of life. Since the complete response or remission (not cure) rate was in the 70 percent range, most patients left the hospital in much better health and spirits than when they came in. It was extremely rewarding to be one of the prime players in this melodrama. The overwhelming gratitude and appreciation of the patients and their families was the invaluable bonus for a job well done. I knew then that this field, although fraught with death, truly offered a great deal of hope. I wanted to be part of this important work.

It would not be long after I had chosen my subspecialties that Bob too chose hematology and oncology. This pleased me because it meant we would be sharing even more closely the intricate facets of our careers. Although he eventually would establish his practice in Ithaca, New York, the miles couldn't separate our friendship. Even today, his family and mine continue to see each other on personal visits or medical meetings. These kinds of relationships have done much to engender my great debt to the field of medicine.

The years spent at Memorial were exciting ones for Joan and me, but they were also intense. I was challenged by a diverse range of tumor types as I became acquainted with hundreds of cancer patients. I learned a tremendous amount from the physicians at Memorial whose medical expertise I came to admire and wished to emulate. The long hours I spent with medical texts and in the library learning about the illnesses that I was seeing everyday gave me a strong foundation on which to build for the years ahead. It became obvious that the study of cancer was a volatile one. Cancer therapy was always changing. Improved methods made old treatments obsolete. Statistics amassed on patient-response to new protocols were often disappointing, but sometimes encouraging and I was caught up in a burning desire to know more.

My off hours when not at the library or reading were mostly spent with Joan. I needed recreation and relaxation to relieve the tensions of work and wanted to make the most of what little time I had. We took advantage of the change of scenery and enjoyed a different side of Manhattan during the week nights, frequently eating at nice cafés and restaurants. Weekends off were spent going to movies, ballet, theater, while Sundays were often spent at leisure around our apartment or in various city haunts.

Back on the job, I gradually became aware that the real lessons I was learning, despite all the book knowledge and clinical skills I acquired, were from the patients themselves. I noticed how they reacted to their diseases, how treatment affected their emotions and elicited heroic qualities of the human spirit, and these lessons I began to value the most. Still, I had more to learn about the human condition. At Memorial I was only skimming the surface. I realized much later that it did not prepare me for the real lifelong experience of taking care of someone from diagnosis to grave. The emotional roller coaster on which a cancer patient takes his doctor over long periods of time can be the most exciting, scary, frequently sad, often heartbreaking, but also rewarding experience.

During my elective on the Leukemia Service at Memorial, I met Mike Kruse. He was the fellow on Tim Gi's service at that time, and he impressed me with his intelligence, energy and amiability. Two years ahead of me in his training, he was on the verge of making a decision about where he would go after Memorial. Ultimately, he decided to head into a private practice on Long Island once his fellowship was over. His decision planted early thoughts in my head of going into a private practice in the metropolitan area, but I still had to complete my fellowship year. I was also debating about whether I wanted to go into a practice on the West Coast. Finally, when his fellowship ended, Mike headed east just as he planned.

Joan became pregnant with our first daughter, Allison, in the fall of my final, fellowship year. I felt mostly positive about the pregnancy since we had tried for nearly a year and had been through the early part of an infertility work up. The news of her pregnancy, however, was somewhat disconcerting in that I felt the impact of a new lifelong commitment—a child. It was time to make final plans for all our futures, final plans that hinged on my career choices.

With my years of training nearly completed, I reaffirmed my decision that private practice in my subspecialties of hematology and medical oncology would be the way I could best make my contributions to society. Yet, I realized that if I chose to practice alone, it would be emotionally and physically difficult to deal with very sick or dying patients on a daily basis. Since my sympathy for these patients and their families and my own personal hopes were intrinsic in my desire to care for them, I would burn out too quickly if I practiced solo and dealt with perpetual disappointments and frustrations alone. From my observations of other oncologists during my fellowship, I realized how vitally important it was to protect my emotional freshness. After much consideration, I concluded there would be safety in a group practice. I would be able to share the responsibilities and the hardships with other equally qualified physicians, safeguard my emotional involvement with my patients, and thereby remain more effective as a caring physician.

It was almost two years after his departure from Memorial that I met up again with Mike Kruse. He had a busy oncology practice on central Long Island's North Shore. We spoke about his need for more coverage and mine for a job. He and another doctor, Helen Delta, who had been a hematologist in the community for seven years, were covering each other on weekends and would be interested in joining forces. Mike invited Joan and me out both to have a look at the area and to discuss the possibilities of a partnership with Helen and him.

When Joan and I came out to visit we were impressed with the natural beauty of this part of the island, so different from the densely suburban, geographically flat south shore in Nassau where I was raised. Here the villages and hamlets retained their characteristic quaintness, yet were somewhat modernized without being faddish. There was space to breath and stretch without feeling isolated. Rolling hills and lush greenery seemed everywhere.

As Joan and I saw it, the community hospitals were well equipped, and Mike and Helen's professionalism complemented the entire picture. It was too attractive an opportunity and community to cast aside. After some discussion, Mike, Helen, and I decided to start a medical partnership and serve the four local community hospitals as hematology and oncology consultants. At the same time, we would take primary care of patients who needed our expertise. All this was to begin in mid July just before the baby's due date and about two weeks after I completed my fellowship at Memorial.

Joan and I were hoping we could squeeze in a vacation break before I began the new practice. It was a tight schedule, but we thought we could do it. At the end of June, Joan quit working for the plastic surgeons and we rented a home in Stony Brook for the first of July. After quickly settling in with only the important boxes unpacked, Joan and I began our much needed vacation. As luck would have it, we didn't get the full two weeks.

Allison was born on July 13, 1975. Joan's early labor was, we feel, induced by a fabulous poker hand. We were on vacation at Fire Island with my sister and brother-in-law, and given Joan's condition, poker seemed like a relatively safe activity. During one particular hand, the game became extra exciting as the stakes grew. The real contest, however, developed between Joan and my sister, Fran. Each time Fran bet, Joan raised her. Straight faced, the two of them kept raising each other until finally, with the pile of chips heaped in the center of the table, Joan called. Proudly displaying a full house, Fran assumed the pot was hers and moved to rake in the chips. Giggling, Joan stopped her and presented four eights! We all roared with hilarity. This howling laughter when Joan showed her hand (we thought later) may have caused her water to break that night, cutting our vacation short to begin our family.

"She's pink as bubble gum and looks just like me," I told Joan after she awoke from the general anesthesia she received for her Caesarean section. I had just come back from the Pediatric I.C.U. where Allison was being kept for observation. They were concerned about infection because Joan had prematurely ruptured her membranes. However, these concerns could not cloud my sense of joy and wonder. This feeling that your child is an extension of yourself is at least part of why parents love their children, and I was no exception.

Two days later, I started working with Helen and Mike. I discovered the tremendous need for our subspecialty then as our workload grew rapidly. Each of us worked long hours either rounding at one hospital and then going to the office the rest of the day or rounding at three hospitals and doing all the hospital consultations.

Life was quite hectic in this transitional time but we all managed. Leaving Joan and Ali in the morning, working all day in a new job, and coming home in the evening to a very tired wife, who was still recovering from her C-section, proved very different from our former life.

Joan's mother, Betty, stayed for three weeks during this time and selflessly helped us through the difficulties of a new baby and a new house. Whenever we needed her, she would take care of her grandchild (eventually grandchildren) with such love and devotion that she put all our worries to rest. Later, on occasion, she would encourage us to take brief vacations away from our home responsibilities because she knew we sorely needed to rejuvenate ourselves and our marriage. It's no wonder that, over the years, I have grown to love her like my own mother.

As time passed in the practice, Helen, Mike, and I found that the patient load had increased to the point that it was taking its physical toll on us. Long days and working into the early evenings, day after day, week after week, made us feel like automatons. We found that we did not have the reserve to give that extra effort that was often required to make our patients feel they were getting individualized and compassionate care. It was time to bring another person into the practice.

Sam Weis filled this position splendidly. He brought a combination of kindness, patience, and a relaxed attitude, a laissez-faire, that fit well into the group. His background and training, including a Ph.D. in biochemistry, added to our professional identity.

Two years later, at the same time that we again felt the need for more manpower, Hank Holt, who had been practicing in the area for a few years, was feeling the rigors of maintaining his own private practice. We were already impressed with his knowledge and expertise that had been demonstrated at the tumor conferences we attended, and felt that he would be a fine addition to our group. Hank agreed that the union would benefit us all and merged his practice with ours.

Also, at this time in my personal life, we had another welcomed addition to our home.

"New toes!" Allison exclaimed when we brought Jessica home from the hospital. The new baby's tiny feet, so pink and wrinkled, astounded our toddler as she watched me take the infant in my arms. Our second daughter increased our happiness as I relearned the joys of family and fatherhood. I looked forward to watching and participating in the development of another child. Some of the anxiety and tension that frequently accompanies a first child were gone now, and Joan and I found we were better able to enjoy Jessica for it. Soon after Jessica was born, we made another and final move to our present home.

Moving the two small offices our group shared to larger office suites in two medical complexes was a complicated effort that took place about this time as well. We hoped these larger accommodations would be more pleasant for the increasing numbers of patients in our waiting rooms. More space would also allow us to continue seeing the patients at the same time without tripping over each other. In addition, we had more room to set up labs and x-ray facilities right on the premises. This, because it reduced the processing time of blood work and film results, served us and our patients more efficiently. By doubling the numbers of nurses, technicians, and staff, and hiring a social-worker therapist to work with the patients and their families, we were getting closer to fulfilling a personal dream of mine: a cancer center providing our patients with total healthcare.

Even after all these changes, the demands of patient care were greater than our combined efforts could satisfy. Although it was obvious that we would need to consider finding another partner in the not too distant future, for the time being we allowed the five of us to carry the workload as best we could. The possibility of a sixth partner was temporarily on hold.

Workload was not our only problem at this time. The profession itself was changing, bowing to the increasing demands of federal and state regulations that were trying to curb costs. It was insidious at first, a little rule here, another regulation there, didn't seem to be such a threat to the autonomy doctors had in the healthcare of their patients. Everyone agreed costs needed to be controlled, and doctors were cautiously trying to do it while still delivering the best care to their patients.

By 1983, not yet ten years after our partnership began, the federal government took aggressive steps and drastically changed the practice of medicine as we had known it up to then. It established the D.R.G's, Diagnostic Related Groups, which for all intents and purposes categorized all known diagnoses by subtypes (approximately 450 diagnoses in all) and gave each a price tag based upon the average expected cost for a hospital stay according to region. This determined how much Medicare would pay regardless of what the actual costs were for medical treatment.

In an effort to make the practice of medicine more cost effective, the government made the actual hands-on treatment of patients more complicated for all doctors. Making medical decisions about our patients' treatments was now limited by the tightening purse strings of the government and insurance companies. This control eroded our authority as doctors and raised the question: How are we supposed to practice now?

Times were certainly changing everywhere. At home, Joan and I no longer had those carefree moments to spend exclusively with each other. As our daughters grew, they demanded more from us. Adult conversation for any length of time became difficult and privacy was sacred. Despite all these hassles, however, I felt blessed that we were dealing with problems no weightier than what to feed the children when they grew finicky or how to handle an overtired, cranky kid. Eventually, as our daughters got older, we enjoyed going on vacations together, spending the quality time we needed to learn more about each other in an environment we picked to be fun and stress-free.

* * * *

Over the years, during such changing times, my philosophy for living life to its fullest continues to develop. It includes the love I feel for my wife, family, and friends and encompasses my need to care for and about my patients and their families.

Joan and I have learned to depend on one another and it is a special secure feeling for me to know she will be there when I need her. This bond of love and caring that we share with each other and our children has always been one of the moorings that I can return to when life seems difficult. It has helped me cope not only with the stress of my practice but with the questions and fears of my own mortality that I see mirrored in my patient's faces.

Many physicians deny their own mortality and feel that they are invincible. This may be a necessary defense, enabling them to deal effectively with catastrophic illness. Sometimes, this denial of their own mortality, however, can act to separate them from their patients. This separation can spill over into their family life as well, making closeness to family members difficult.

I too have struggled with these issues and have come to an understanding. People are better able to give and receive love when they know the appropriate time to let go of the beloved. Sometimes, that letting go is as important as allowing your child to grow up and become an individual or to get married. Other times, it is as sorrowful as letting an aging parent or sick spouse die, loving them enough to free them from a life of torment. Likewise, physicians are better able to render kind and compassionate care only if they can reconcile attachment with detachment and not feel prolonged pain after the deaths of most patients. For me personally, letting go is an important part of being a good parent, son, husband, and oncologist.

Fear of one's own death is also inherently involved in an oncologist's ability to relate to suffering cancer patients. If a doctor suppresses fear about his or her own mortality most of the time, that doctor may find it too difficult or even too frightening to comprehend what their patients are going through.

Recognizing and accepting, albeit with fear, the possibility of my own death allows me to better appreciate and celebrate life, deal effectively with people facing death themselves, and ultimately allows the letting go of myself required, at least on some level, for loving.

3~

The Courage To Change

"Doctor Berger," my secretary called over from her desk, "There's a Doctor Cohen on line 3 for you. He says he has a patient he would like you to see."

"Okay, I got it," I said as I dodged one of my nurses who held a syringe full of chemotherapy and picked up the phone in the nurses station.

Despite the mild February weather outside, inside the office, it was an extremely hectic Wednesday, and I was under my usual self-induced pressure to try and stay on schedule. Recalling my credo (Cancer patients should not waste their valuable time waiting in a doctor's office) urged me to hurry. But, on this particular day, I was feeling lots of anxiety over any delays. Doctor Cohen's call couldn't have come at a busier time.

"Hello, this is Doctor Berger," I answered.

"Good morning, Doctor Berger. My name is Sam Cohen. I'm a general surgeon in Patchogue and I'm calling to ask you to see a patient on whom we just biopsied a supraclavicular lymph node. Pathology is definitely calling this a malignant lymphoma. But whether it's Hodgkins or another aggressive lymphoma is still in question. On the recommendation of the patient's relative who has a medical background, there are going to be three pathology reports to determine subtypes, but so far, the first report has concluded that the malignancy may be difficult to type."

"How old is the patient?" I queried.

"She's in her mid or late twenties. She originally presented with chest pain and her internist took a chest x-ray which showed rather extensive mediastinal adenopathy..."

As Doctor Cohen spoke, my mind began to race ahead. That morning I had already seen about ten patients. Some of them had completed their treatments— mostly adjuvant chemotherapy to help prevent cancer recurrence. The majority were being seen for incurable cancers for which we could only hope to have a transient period of control, a control longer for some than others. This new case

37

could be different. This young woman sounded as though she could have a curable disease.

My body tensed with excitement. Curable cases required the most exacting judgment. The challenge was that they required a fine tuning of protocols, unlike the incurable cases. I would need to work much harder, to rely more heavily on all my medical and oncologic training and expertise gained over the years. The burden of responsibility would lie squarely on my shoulders. And if I cured her, I would feel as though I had made not only a meaningful difference in the life of this young woman, but also a great contribution to mankind as well. My craft and art as an oncologist would prove that medical science could at times vanquish cancer. The important role I would play in such a scenario, I suppose, is what had attracted me to this field.

At the same time, my excitement over her potential long-term remission and even potential cure was counterbalanced by the total sense of devastation I would feel if there were poor response by the patient, a relapse, and then death. It would be a tragic end for a young, viable life.

"I appreciate your calling, Doctor Cohen. If you give me her name and number I will call her today and begin our evaluation." I replied after he finished with the details of her case.

He sounded relieved when he gave me her name, Lynn Moline, her phone number, and promised to send me copies of her records in the next few days Thanking him, I assured him we'd keep him informed, then hung up.

Before I could hurry away to my waiting patients, I first asked my secretary to schedule Lynn Moline for admittance to the hospital the next day during my rounds and to remind me to call her at my lunch break. Then I picked up a file and went off to find my next patient.

Later, at lunch, there were numerous calls to make and unattended paperwork piled high on my desk. As I quickly ate the usual chicken with broccoli, I buzzed my secretary, asking her to call Lynn. I had just finished the last mouthful when the intercom interrupted, "Doctor Berger, I have Lynn Moline on line 4."

"Thanks, Chris." I pushed line 4 after clearing my mouth with a quick sip of a diet soft drink. "Mrs. Moline? Hi, I'm Doctor Berger." I began pleasantly, "Doctor Cohen asked me to call you and set up an appointment to start a treatment plan..." Somehow, in a brief phone call, I would have to communicate my genuine concern to this young woman about her condition, assure her that I would be able to help, and still be succinct. I did not like the idea of shortchanging her at a traumatic moment in her life. Tomorrow, I reminded myself, when I see her at the hospital I would explain everything in detail. Although my words were polite, I felt tense. The morning's hassles mixed with stress, anxiety, frustration, and my gobbled lunch had done a good job of constricting my colon.

* * * *

The next morning, I started hospital rounds early and was nearly finished with the third floor when the floor secretary informed me that Lynn Moline had just been admitted to room 314. I thanked her, picked up Lynn's empty chart, and walked toward her room.

Lynn was an attractive brunette. She had just changed into a nightgown and bathrobe. Her husband, a pensive-looking man, appeared anxious as he sat in the chair next to her bed.

"Hi, Lynn. I'm Doctor Berger."

She smiled warmly and extended her hand which I took. Reading a patient's body language was important in helping me understand what kind of person I was going to be dealing with emotionally: Lynn's handshake was firm yet not overbearing, as though she, under ordinary circumstances, was accustomed to being self-reliant. She also seemed to be in good humor, without pretense. If these were early indications of her actual personality, she might turn out to be a helpful, pleasant patient. "Hello, Doctor Berger," she replied. "It's nice to have a face to go along with a voice. This is my husband, Russ."

"Nice to meet you, Russ." I said as he rose politely to greet me. I could see Russ's face was sad and drawn, unlike the face of his smiling wife. He seemed to be in worse shape emotionally than she was. "What I'd like to do is sit down and take your history, examine you, then give you an idea of what we would like to accomplish while you are in the hospital. If you have any questions after that I'll be happy to answer them, okay?"

"Fine." Lynn said pleasantly. Russ nodded but remained quiet.

It took about a half hour to get the details of her presenting complaints and how she had gotten to this point. Her medical care by the South Shore doctors had been quite adequate, but some of the information she provided seemed more technical than usual for a patient.

When I asked her if she was familiar with medical research, she smiled, "I guess you might get that impression, but no, I'm only learning about my particular illness as we go along. Actually, Russ's uncle is a cytologist, and he has advised us on what to do so far. Already he's had my slides sent for review by some of his friends who are expert pathologists on Long Island."

"Uncle Rich," Russ broke in for the first time, "is also recommending that Lynn's pathology slides be sent to two well-known lymph node specialists on the West Coast." Despite his reserved manner, I could see Russ was keenly involved in the plans for Lynn's healthcare.

"I can't remember their names. If you want," Lynn suggested, "I can call Uncle Rich and get you their names and phone numbers. I think they're at Stanford."

"Any information you can give me would be very helpful and important. Please do that as soon as you can." I said as I wrote a note to myself in her chart about the pending specialists' reports. "Do you have any medical records with you now?"

"Oh, I thought Doctor Cohen was sending you copies of everything," Lynn replied, "but I do have a copy of one of the pathology reports; this one's from the hospital where I had the biopsy. My uncle suggested we should get our own personal copies of the reports." Lynn paused and then chuckled, as she added. "Interpreting them is another matter, but it's been great having Uncle Rich to help us." Lynn began searching through her purse on the bed stand. "Here it is," she handed me the report.

Lynn was demonstrating to me the desire to participate in a positive and helpful way in her own care. This would be a major asset, I felt, and was pleased to see it.

Perusing the report, I read the words malignant lymphoma. The text went on describing the findings that leaned toward Hodgkin's disease, but then other architectural descriptions would lean toward NonHodgkin's Lymphoma. It was clear, as Doctor Cohen had mentioned the previous day, that the pathologist was not sure. Since treatment of the two disorders differed, it would be important to get an expert opinion. The pathologist noted that he had sent the slides to a lymph node pathologist at Brookhaven Labs, a man whose reputation I respected.

"Lynn, do you know if there is any word yet from the pathologist at Brookhaven Labs?" I asked.

"We haven't heard anything yet." She shrugged as she glanced at Russ, as if for confirmation. He didn't contradict her, but nodded his agreement.

I could see this case still had lots of loose ends to clarify before a clear-cut treatment plan could be formulated. This would not deter me. It just meant phone calls and some delays with enough of the good, old, bureaucratic hassle-factor for good measure. No matter. I would not let these things get in our way.

Next, I examined Lynn and noted the area of biopsy which was healing well. She had a few other enlarged lymph nodes which were firm and rubbery in her neck. The rest of the exam was normal. She was in good general health, which would help her fight her disease in the months ahead. Even though she was showing signs of worry and concern, her emotional attitude was cheerful, as though she was trying not to let her fears dominate her otherwise sunny disposition.

"How bad is it?" Russ could no longer hold his silence.

"Well, it's too soon to say." I was honest. "There is little doubt that we are dealing with a malignant process involving Lynn's lymphatic system. Our job is to define exactly what the process is and how extensive it is."

"How do we do that?" Lynn asked, her voice was steady but filled with concern.

Although I could sympathize with her fears, I was glad she was using the word "we." I use it from the beginning with patients to emphasize that I, my group, and indeed, all the physicians involved in a patient's care are part of a team. The rest of the team consists of the patient, his or her family, and anyone else they want to be part of it. The battle lines are drawn. It's the team against the disease. The patient is an integral part of the team, the captain. I consider myself the coach. In this way, not only does the patient feel less helpless, but the patient's role is very important in determining the strategy to beat or arrest the disease. Lynn seemed to understand this intuitively.

Answering her question, I explained, "We'll get all the pathology opinions and see if we can come to a conclusion as to the exact kind of lymphoma you have. That may mean sending it out to yet another pathologist if we have to. While we are doing that, we'll put you through a series of tests to determine the extent of the disease. These will include x-rays, CAT scans, bone scan, bone marrow examinations, and a lymphangiogram..." I enumerated and then explained each test painstakingly so that there would be few surprises for Lynn or Russ.

It was scary enough going through the process. The less that remained unknown the more comfortable they would feel. This was part of what a good and concerned oncologist could do for his patients. Unfortunately, it was frequently time consuming, and sometimes constraints made me feel that I might not have covered everything 100 percent. With the Moline couple, I took special care now so that all questions would be answered to their satisfaction. I also needed to know that the Molines were comfortable with my care. We had a very rocky road ahead and the trust that I engendered now would serve us in good stead later on.

"How long will all this take?" Russ asked dryly. As with many spouses of cancer patients, he was quite fearful about Lynn's condition.

"About five days," I replied.

"Then what?" both he and Lynn said simultaneously.

We all smiled.

"By then we should be able to decide on a treatment regimen." I didn't have the time to go into all the details about therapy especially since a lot depended upon Lynn's pathology and extent of disease, so I added, "I'll be back next week to go over everything with you. One of my partners will see you every day until then to give you the results of the daily tests and answer your questions. You will have the benefit of our combined medical expertise in formulating your treatment."

"Why is it?" Lynn hesitated noticeably. "I've always watched myself carefully. I don't smoke, I eat healthy foods, I exercise, I take vitamins, I don't even drink. What did I do wrong? I mean, to get ... cancer?" Her face had grown serious but she still seemed controlled emotionally.

My heart went out to her. Not only was she bearing the brunt of this terrible diagnosis, she was also taking the blame for it.

"Lynn, you have enough to be concerned about. There is no blame here. The fact is no one really knows what causes this disease. What we do know is that if you have to fight cancer, this is one of the most treatable and sometimes curable forms of it. I'll be better able to tell you much more about that at the end of the week."

"So, I have a lymphoma that can be cured?" Lynn asked positively, focusing on the most important information she had just heard.

"Yes." I answered though I did not want to frighten her with the other half of the truth that the aggressive lymphomas kill more rapidly than their less aggressive counterparts. Unless chemotherapy worked, she would not have much time. "We know that your lymphoma is one of the swift-acting kinds. We'll begin chemotherapy as soon as the workup is complete." I added in all honesty.

"I'm ready to do whatever needs to be done," Lynn said with resolve.

"I felt that within minutes of meeting you," I replied congenially. "I know I can depend on your cooperation, and you can depend on me and my group to do everything we can. With a team like that, how can we lose?" I felt we were near ending this first meeting on a very positive note.

"Before I go, do you have any more questions?"

"Yes. One last question." Lynn said very softly with a pained look on her face. Her voice was on the verge of trembling. "Coming from a large family myself, I always assumed I would be having children," she said dolefully looking down at the foot of the bed so as to avoid any eye contact with either me or Russ. "In fact, when I first began to feel sick, I refused x-rays because there was a possibility that I could have been pregnant. Now, what will chemotherapy do to me and my ability to have children?"

I was surprised by her question. Apparently, as potentially life-threatened as she was with the lymphoma, she seemed less disturbed by it than she was about this problem.

"How old are you?" I asked.

"Thirty. I just turned thirty." She answered.

"Ovarian function is frequently affected by this aggressive form of treatment. Sterility is a possibility and often occurs in patients who are well into their thirties and older." For the first time, I could see tears in her eyes as I answered her. Obviously, this was very important to her for her to display such emotion, especially since she had controlled herself so well up until now. I tried to ameliorate her pain, but at the same time, keep reality in my forecast.

"Lynn, I can't deny the possibility that this treatment may cause permanent sterility. Let's try to keep things in perspective. Right now we are trying to cure you of a disease that has the potential to kill you. If you can't have children as a result of the treatment, we will all be disappointed, but in my opinion, that is a price worth paying if it means effecting a cure. The good news is that a high percentage of women under the age of thirty recover their ovarian function and go on to be able to have normal children."

Lynn dried her eyes with a tissue that Russ offered her and apologetically said, "I'll be all right. Of course I have no choice. I had been worried about becoming sterile, but finally hearing it as a real potential just threw me. I'm sorry for this outburst."

"It's about time," I assured her. "You've taken all this in like a trooper. You're allowed to show some emotion. As a matter of fact, it's healthier to let some of the anger and fear out, rather than to hold it all in. Anytime you're angry or upset, it's okay to show it. We'll all deal with it together. Nobody said this is going to be easy. Good things rarely come easy."

She glanced briefly at her husband and then managed a frail smile behind her tears. "If sterility should happen to me, should I assume that I'd need some hormone treatments or... or I might grow a beard?" The funny question coupled by the switch in her voice indicated that she was joking to lift away her concern. Yet, her humorous attempt to mask her worry did not lift the pallor that had settled over Russ's face. Tight-lipped, he was dealing with the information by suppressing any reaction.

Because he didn't respond, I decided to play along. "Not as nice as mine." I said while stroking my own full beard. I was glad and relieved to hear her laugh genuinely at my reply. Even Russ was amused. "Honestly," I advised in all seriousness after their chuckles subsided, "Since your age puts you right in the middle, we cannot say one way or another how chemotherapy will affect your reproduction. Let's just work on your cure first, and then worry about progeny after."

Lynn smiled. She looked at Russ, they both looked at me. We all knew we had just jumped over a major hurdle together.

* * * *

Five days passed quickly. I transiently thought about Lynn. There were too many other patients, meetings, and involvements for me to spend any more time on her case. As long as she was on track for her workup, I knew I could pull it all together when I saw her next.

When I entered St. John's Hospital early the following Tuesday for morning rounds, I knew I needed to start with Lynn. I reviewed her chart and was glad to see her disease appeared to be limited to her neck and chest. I was particularly happy that her bone marrow biopsies were negative for tumor. This meant that we could potentially use bone marrow transplantation as part of Lynn's therapy, if need be. Then I darted into Lynn's room.

"Good morning," I hailed. "Before we start discussing the details of where we are and where we're going, Lynn, have you heard anything from your uncle about the other pathology reports?"

Lynn seemed startled by my sudden appearance, but answered quickly without missing a beat. "Just yesterday, I received some items from my uncle. Another pathology report along with the names and addresses of the west coast specialists at Stanford." She rummaged through her belongings and brought out a business size envelope. "This one's from Brookhaven Labs."

"Great!" I took the report to copy it and headed out of the room again, promising to return shortly. The two pathologists at Stanford University were indeed well known hematopathologists, and I would be eager to receive their reports when they arrived. In the meantime the pathologist at Brookhaven commented on how difficult the case had been and concluded that this was an intermediate to aggressive NonHodgkin's Lymphoma. Because he could not be absolutely certain, it was becoming obvious an arbiter would be necessary to help us decide on a treatment plan.

So as not to lose any more time, I thumbed through my phone book and found Doctor Ben Singer's number at Memorial Sloan-Kettering. Ben and I had been oncology fellows together at Memorial. He stayed on as a full-time attending on the lymphoma service, and had been invaluable to me when I had difficult cases. We discussed Lynn's case and it quickly became clear to him as it was to me that a full review of the entire case with all x-rays and especially the slides would be the way to go before starting treatment. A second biopsy might be necessary.

"I planned to start her on treatment today," I confided, "so I feel this is a setback."

"If she can be admitted here some time this afternoon," Ben accommodated, "I'll arrange to see her this evening."

"Excellent," I cheered, then thanked him and returned to Lynn's room, relieved that I was able to make all the contacts easily and to achieve an agreed-upon next step. Now I had to convince Lynn, which I was confident would not be difficult.

"Sorry I was a little abrupt before, but when I'm on a mission, I sometimes lose track of the everyday niceties."

"I understand," she smiled.

"Well, I'm sure my partners have been keeping you informed of the results of all your tests as they have come back."

"Yes," she nodded. "They've all been very nice. And the news is not totally bleak either."

"Everything's back and I'm glad to say your bone marrow is negative for lymphoma and the disease appears to be localized above the diaphragm."

"What does that mean?" Lynn never seemed too embarrassed to admit to ignorance when it came to medical terminology. Her curiosity showed me she was obviously willing to learn.

"Well, that puts you at a stage II, which is better than stage IV. Occasionally we have been able to treat and even cure people at the worst stages, but stage II certainly is very treatable and possibly curable." I began using that word both because it was potentially true and Lynn needed, I felt, an infusion of hope at this point. I knew she would do whatever needed to be done to live. She would fight very hard and I wanted to give her further impetus to fight for her life.

"When do we begin treatment?" she asked quietly.

"We've got a bit of a problem with the pathology." I answered honestly. "I'm still not clear as to the exact type of lymphoma you have, Lynn. So far, the pathologists differ in their opinions. A second biopsy might even be necessary. I took the liberty of calling Sloan-Kettering and getting you admitted immediately, if you can get there by this afternoon." Lynn was silent, but attentive, so I continued answering the questions I presumed she had. "Doctor Ben Singer and his group treat only lymphoma patients. We can accomplish two things by going there. First, they will review your entire case and review the slides and help render a decision on your histology once we receive the final pathologist's report. Also, they can easily schedule you for a biopsy if it becomes necessary. I especially need to know if it is Hodgkin's versus NonHodgkin's, because the treatment does differ. While you're there, you can be considered for bone marrow cryopreservation."

"All this sounds awfully serious," Lynn commented wryly.

"It's important to act fast on this. Now bone marrow transplantation, which is a relatively new and exciting technique, enables us to give very high dose therapy to patients with malignancies, something we could not do prior to this technique. We take out enough bone marrow that is presumably void of cancer cells, give the patients a dose of radiation and/or chemo that would ordinarily wipe out the marrow and, we hope, all the cancer cells in the body, and then salvage the patients by giving them back their untreated bone marrow cells."

"I was hoping I could have my treatments here," Lynn admitted softly. Her impassive face and steady voice misled me at that time into believing she was totally unafraid. It would be months later, after I had gotten to her know her better, that I'd be able to interpret her quiet manner as her reserve way of expressing emotion or fear.

"First, we must be as sure as we can about your diagnosis before undertaking therapy." I replied, unaware of her subtle struggle. "I am confident that you can be treated here, afterwards."

"You would still be taking care of me then?" Apparently within a week's time, she had put her utmost faith in me.

"Yes, when you're on the Island. I will also be in constant touch with the lymphoma group at Sloan-Kettering. I'm not giving up on you, I'm trying to give you the best shot at cure. Don't worry, we're still a team." I reminded her, as I gave her a gentle pat on the shoulder.

"Whatever you recommend, Doctor Berger." she agreed softly.

I was pleased with Lynn's ease of understanding and apparently predictable response. This was never easy, but she was making the path as clear of emotional obstacles as it could be.

Immediately, I arranged for her chart to be copied, all her x-rays to be made available, and a copy of her slides to be sent to Ben. Then I discharged Lynn and told her to let me know as soon as she returned from Memorial.

* * * *

"Hi, Ben," I placed Lynn's chart in front of me as soon my secretary had buzzed to tell me he was on the line. "How are things in Mecca?" I joked.

"Easier, I wish," he retorted cynically. "We reviewed Lynn Moline's workup and slides yesterday when she arrived and she has just undergone another lymph node biopsy this morning. We agree with your staging but we were having a real hard time with the pathology. Our group needed more fresh tissue. The pathologists felt they could make a more definitive decision with special stains and monoclonal markers."

"How did Lynn respond to the second biopsy decision?"

"She was agreeable. Although she said she'll do whatever is necessary as long as it's okay with you," Ben replied.

"Did you tell her we had agreed upon this yesterday?"

"Yes. She was assured it was your decision as well. I'll let you know as soon as we have some more information. Good-bye, Roy."

"Very good. Thanks, Ben, for keeping the ball rolling."

I was happy that I had sent Lynn to Memorial. This way, she and her family felt they were getting the expertise of a world-renowned cancer care institution, and I was getting the valuable input I needed to help make a critically important decision regarding her treatment. This sharing of responsibility was reassuring both to me, and I was sure, to the Molines. If the end result of Lynn's case was poor, everybody involved would know it would not have been for lack of medical expertise or attention.

Another week went by when I received a call from Ben. This time, his greeting was curt, businesslike. "Good morning, Roy. The result of Lynn Moline's second node biopsy showed no evidence of disease. But, in our opinion, the first slides lean toward a mixed histiocytic-lymphocytic NonHodgkin's lymphoma. The results from the west coast pathologists are still not in."

"Okay." I paused. "What do you recommend?"

"We feel we have enough to go on, although her family seems eager to get the results from the two specialists at Stanford. Right now, we would like to place her on our current program of treatment with the L-17M and later randomize her for possible bone marrow transplantation."

I was not surprised by his recommendation. The L-17M was a specific chemotherapy program, using different drugs, which was designed to maximize cell kill in order to effect a remission. But I immediately became concerned that the random selection by computer could place her in the transplant arm of the protocol. At that time, research being conducted on the total experience of marrow transplants for lymphoma was still in the early stages. The course was fraught with severe toxicity and a reasonably high mortality rate. "When does the randomization take place?" I queried.

"After the initial three courses of the induction of the L-17M," Ben responded.

"Have you spoken to Lynn about it yet?"

"Yes, she's willing to start the program but does not want to commit to randomization yet."

"Is that a problem from your end?" I asked.

"You know we need answers for these diseases!" Ben snapped peevishly as if this were an old argument he had had before. "Our protocol is designed to help us understand if transplantation plays an important part in the treatment of this disease," He sighed and then slowly continued with less of an edge to his voice. "That shouldn't stop her from beginning treatment here immediately with us."

"I know." I understood his frustration. "Well, once we begin her treatment, we'll see if and how she responds. Then we can all face the randomization question down the road."

"Fine. I'll start her today. You can arrange to continue her four-week induction course out there, since you know the L-17M, and it will be so much easier for her to be closer to home. " Ben was starting to sound better.

"When would you want to see Lynn again?"

"After she completed the last induction course," Ben returned. "By then, we should have a better feel for how she's doing."

"Okay, Ben, thanks for all your help. At least now things are clearer, and we have a definitive game plan."

"You're welcome. How about sending us an easy one now and then?" Ben joked.

"Ben, we can handle the easy ones. We only need you for the tough cases," I quipped good-naturedly.

He laughed before he replied amicably. "Roy, let me know how she does and if I can do anything else for you."

"Will do. Thanks again," and I hung up.

I began thinking about the randomization arm to bone marrow transplantation. It was extremely dangerous and I was not sure it was needed. I had seen patients do very well using the L-17M alone. Would it really be necessary to add transplantation and all its risks if Lynn responded to the L-17M? Well, I was relieved that this question did not have to be answered now. First we needed to see if Lynn would respond to the first phase of the L-17M.

After Memorial administered her first chemo, they discharged her and sent her home. Her treatment was now my responsibility.

While "induction" did not have the very high doses associated with bone marrow transplantation, this regimen of chemotherapy was not going to be easy. It was designed to deliver intensive antineoplastic treatment to all tumor cells which were generally more sensitive to chemo than normal cells. Since it was delivered via the bloodstream, every cell in the body, including the normal ones, was exposed to it and even the normal ones were affected to some degree by its toxicity. At first, Lynn was able to tolerate the drugs with minimal side effects, some nausea, vomiting, with the most obvious effect being her hair loss, which began a week to the day after her first chemo. Despite the transformations chemo made on her body, Lynn was a cooperative, pleasant, and intelligent patient who maintained a very positive attitude.

Throughout March, I got to know Lynn better when she came regularly to the office to continue the induction course which Memorial had begun. I began to admire Lynn's courage and her determination. She, in turn, liked me, appreciated my honesty, and trusted my expertise, but I realized it was not a blind trust. With a keen desire to know and understand everything, she politely questioned me and my staff about what her blood counts were each day, what the x-rays showed, what each test result meant, and how these results applied to her own progress.

While she listened and understood the gravity of her lymphoma, she always seemed to keep her emotions in check. I appreciated her control. It made her easy to treat because she didn't seem to require false assurances or constant emotional support. It was not necessary to cajole her or to use stern reminders of the gravity of her illness to keep her on track with therapy. Rather, Lynn was always ready to do whatever was necessary whenever it was needed. She truly seemed to understand that dealing with the disease was her first and foremost priority.

Observing her, I wondered if I could keep up the same positive attitude and tremendous fortitude in the face of an aggressive malignancy like hers. She, of course, never had the exposure of watching someone die of malignant lymphoma, and so perhaps her ignorance was working in her favor. She freely admitted that her optimism was a form of deliberate denial. She chose to focus mostly on the positive, to dismiss the negative. She refused to lose sight of her goal to be cured. Her religious upbringing combined with her own personal philosophy had given her a tremendous faith in her life, her purpose for living. "It's not my time," she said softly and smiled. I knew she believed this without question.

Two weeks after her discharge from Memorial, she was ready for her first post-induction office visit. I was very anxious to examine her and see if she had any response to treatment, but before I could ask her to sit on the examining table, she began reporting on what she thought were unusual side effects. "I expected to loose my hair and feel lousy and to have bouts of nausea, but why has my appetite increased to the point where I'm eating all day and still am not satisfied?"

"That's the prednisone." I explained. "It's a steroid which has a direct antitumor effect on your lymphoma and among other things, it stimulates your appetite. It can also give you a sense of well being, make you hyperactive, cause insomnia, and give you water retention." Reciting the list of side effects was a regular routine for me.

"That explains why my face looks so full and bloated and why I'm gaining so much weight," she replied good-naturedly patting her moon face, "especially from all this eating. The cravings are unbelievable, too." she continued. "Lately, at odd hours of the night I've been waking up desperately hungry for lox...sliced thin, with cream cheese, onion, and tomato on a warm bagel." She looked hungry.

As I approached her to examine her, I leaned forward in a feigned tone of confidentiality and commented matter-of-factly. "That's another side effect of the prednisone, Lynn. It makes you Jewish." My comic delivery took her by surprise. When she giggled in response, I cracked a reciprocal grin. It was nice to share a bonding moment of levity.

There was more to be happy about. As I examined Lynn, I was elated to find that all the enlarged firm, rubbery, malignant lymph nodes in her neck had totally disappeared. There was no doubt in my mind that this was due to the first L-17M induction course she had just completed.

She looked at me with an expectant and quizzical expression. The unstated question did not need to be asked. "All your nodes are gone," I smiled triumphantly! "Let's take a chest x-ray and see if they're gone there too."

Lynn was beaming ear to ear as was I. I knew that her chest film would confirm my physical findings. But, the disease in her chest had been very bulky, so I didn't know yet if it would be completely gone there. Soon, however, we were both looking at a normal chest x-ray, alongside her pretreatment film. The difference was glaringly obvious even to an untrained eye. The first cycle of treatment was a great triumph. It was clear that we had picked a treatment that was so successful that it put Lynn's disease into a complete remission within two weeks.

"This is wonderful!" Lynn said softly with glazed eyes. "I don't know how to thank you..."

"You don't need to thank me," I responded. "Your face has already said it all. And the response itself makes me so happy that no thanks is necessary, but of course, always welcome," I said putting my hand reassuringly on her shoulder.

In the next breath I said, "Now, back to business. We can't let a little success go to our heads. We still have work to do. We need to schedule your next course of treatment."

"Whatever you feel is necessary," Lynn replied happily.

After she had left the office, I began to think ahead, hoping that the next two induction courses would be tolerated well, but also ready to expect anything. I had seen so many different types of responses. It was very difficult to predict reliably how human protoplasm would behave. Some patients never get any major toxic effects. On the other hand, it is very rare to lose a relatively healthy patient as a result of the treatment.

* * * *

By the beginning of April, Lynn had completed the next two courses of induction. As a result of these treatments, she had become more lethargic. Her otherwise upbeat personality seemed muted by her low blood counts and low energy level. Her chest x-rays, however, remained normal, so I could only buoy her up by smiling and triumphantly announcing that her complete response to chemo was clearly continuing.

The next stage of her therapy came to an abrupt standstill. It was now a question of whether Lynn would be randomized for the possible bone marrow transplantation at Memorial Sloan-Kettering. Lynn remained uncommitted. The advice of her family was to wait until the pathology reports returned from the west coast specialist before she considered randomization.

Lynn began making weekly trips to Memorial for follow-up visits and briefing on randomization. Once a week, she would come back to my office for a blood count. During this hiatus from treatment, she was making a quick recovery. Gradually, the unpleasant side effects waned as her blood count returned to a more normal level. Weeks were ticking by. This worried me. Without further treatment, she could relapse.

When, in mid-April, I received copies of the west coast pathologists' reports, I was disappointed. As I feared, they disagreed on the histology of the biopsy. It was certain now that we would never get a precise diagnosis of Lynn's aggressive lymphoma. However, whatever she had was, so far, responding to the decided course of treatment.

One afternoon in early May, I received a call from Lynn's mother. During the entire course of Lynn's illness I had never spoken to anyone from her family except Russ. I knew that there must be an important problem now.

"Hello, this is Doctor Berger," I said.

"Hello, Doctor Berger, my name is MaryRae DeMara. I'm Lynn Moline's mother. I hope you can take a few minutes to speak with me about Lynn," she said.

"Of course, what's the problem?"

"Lynn is very upset about this randomization that they want to do at Memorial. She feels they are pushing her into a treatment that she may not want. I'm calling because she's confused and too upset to talk. She feels that they are not giving her a choice and that if she doesn't accept the randomization, they will not continue her treatments."

"Let's get something straight right away," I said slowly. "Whether Lynn decides to accept randomization or not will in no way stop her treatments on the L-17M. I can tell you that with confidence because it is I, not Memorial, who has been treating her since the beginning, and I have no intention of stopping the treatments if Lynn wants to continue."

"Oh, thank you, Doctor Berger. Lynn will be very reassured to hear that," her mother replied.

"There must be some misunderstanding which I can probably clear up with one or two phone calls, Mrs. DeMara. Memorial may be pushing the randomization process very hard. Their job is to answer the pressing questions for all of us clinicians so that we may know the best methods to treat patients like Lynn. Unless we can compare patients who go on to marrow transplantation against those who stay with conventional therapy, we'll never know if we can improve patient survival with this technique. Since this decision is a difficult one it needs to be done impartially such as by computer selection, not by the patient. That is the reason, I'm sure, Lynn feels she is being pressured."

"Does she have to agree to randomization?" her mother asked.

"Of course not!" I replied, "but bone marrow transplantation may be an important advance. It would not be an arm on the protocol unless there were theoretical and practical reasons why it might be an improvement over standard therapy." I was beginning to run short of time. "I'll tell you what. I'll call Memorial and speak with them to see if we can straighten this all out, okay?"

"That would be very much appreciated, Doctor Berger. Thank you."

"You're welcome. Good-bye, Mrs. DeMara."

Later that day I spoke to Herman Gunan, the physician in charge of bone marrow transplantation.

"Your patient is being uncooperative." Doctor Gunan reported coldly in response to my call. "She's unwilling to be randomized."

"She's been cooperative and reasonable up 'til now." I replied heatedly in Lynn's defense. I knew how some doctors might take a patient's active interest as interference but Lynn had never given me cause to label her uncooperative. "It sounds as though we have a breakdown of communications here. I'm already concerned about this delay in deciding her next phase of treatment—let's not screw up our chances for cure. However, I'm more disturbed by what I've just been told; the patient is under the impression she will be denied treatment unless she is randomized."

Gunan sounded genuinely astonished. "That's unfortunate! We didn't mean to confuse or misguide her. We simply wanted to convince her to accept randomization as an alternative to coming up with this difficult decision by ourselves. We did stress how important it was to abide by the computer's selection especially since randomization is required to make our statistical data relevant."

I realized Gunan's clinical stoicism was the result of his intimate involvement in the program. Perhaps he was taking Lynn's reluctance to be randomized just a little too personally.

"Regardless of your statistics," I continued sternly, "I think you should call and reassure her that whether or not she is randomized, we wouldn't turn her away untreated. Besides, since she's had a good attitude throughout induction I'd like to keep her as optimistic as possible. It certainly couldn't hurt if she felt she had some say in her future treatment." As a reminder, I added frankly, "I find she cooperates when she's treated more like a person and less like a statistic. Try it."

Gunan acquiesced. A day later, I was informed that Lynn had agreed to allow the computer to choose her protocol.

* * * *

"Lynn Moline's disease may be recurring, " Gunan called me late on a Wednesday afternoon at the office.

I was alarmed and startled by his statement. "What have you found?" I asked, carefully controlling the feeling of profound sadness which was erupting deep within me. Standing at the nurses' station, I was surrounded by the usual, steady circulation of staff and patients.

"Nothing definite, yet." Gunan replied. His voice sounded distant. "Except she's complaining of chest pain and shortness of breath which were symptoms when her disease first occurred. Recent chest x-rays show an enlarged heart and the results from the echocardiagram show large amounts of pericardial fluid. It is probably decreasing her cardiac output. Our group discussed this problem and we have already suggested a pericardial window to her for two reasons. It will prevent cardiac failure and also will allow tissue samples to be taken of both the fluid and the pericardium. We'll be able to determine whether the disease has spread by direct extension there."

"Did Lynn agree to your suggestion?" I asked, concerned how this news would affect her.

"Yes, but she wanted to know your opinion on this." Gunan answered stiffly. Obviously, the trust she had placed in me over her care was a sensitive issue with him.

This was not a simple decision. They were talking about opening up her chest, and removing a portion of the pericardial sack which surrounded her heart. It was a big deal. Putting that thought aside for the moment, I asked, "Isn't she due at Memorial to have her marrow harvested now?"

"Yes, that was why she had been admitted. She's scheduled for tomorrow morning." Gunan added, "We'll do the harvest first. We can schedule the window for a week later."

The tension inside me was mounting. I was getting angry. Lynn had been off chemo for the entire month of April. In my opinion, that was four weeks too long. She had been lucky to go this far without recurrence; now, at the beginning of May, her luck might have changed.

Gunan continued, "As you know," he reminded me, "expediency is our utmost concern."

I didn't have to be reminded. I realized that, if the disease had truly recurred, there wouldn't be much time to debate the procedure. It was with a sinking feeling that I quickly agreed to his proposal and allowed Gunan to make the arrangements with the thoracic surgeon at Memorial." Tell her I agree with your decision and please keep me informed," I reminded him in turn.

When he hung up, I felt very discouraged and unhappy for Lynn. The doctors at Memorial were on top of the situation but it was a sad turn of events if she had indeed relapsed. We had seemed so close to cure. This was a case I had hoped would be different. Now I was getting tired of dealing with frequent defeats.

Less than ten years ago, when I did my fellowship at Memorial, I was under a different impression. Fighting cancer was a magnificent undertaking. To a newcomer like myself, the institution's environment—so unified in its purpose—was replete with passionate dedication and adventure. I felt excited working with the powerful chemotherapy drugs, administering to the sick and watching how the skill and expertise of physicians around me were effectively changing the course of the patients' diseases. I was also impressed by the advances in cancer research. With the available funding to back it up, these oncologists were making definite strides in curtailing cancer. At that time, it was like riding the crest of a tremendous wave to shore... exhilarating! However, looking back now, I realize that my early experiences in oncology gave me an unrealistic introduction to actual practice.

The first wrong impression was the false sense of hope I received from seeing only segments of different cases. I started with some patients who were early along in the course of the disease. I diagnosed them, treated them, and they would in turn go into remission. It gave me such a rush, a feeling of power and importance. I was convinced by these apparent successes that this was the field to be in. Rarely did I see these new patients die. I wasn't there long enough.

Other times, I would take over some cases of patients who were at the very end of illness. Since I hadn't known these patients from the beginning, I was able to keep personally uninvolved when they died. Somehow, I had the misconception that I had steeled myself to accept the death of the terminally ill. Emotionally, I had few real defeats.

It was only when I came out to a private practice, and got to know the people and the overall picture of their cancers that I learned the truth. There was a lot of failure in oncology. After I've spent two, three, maybe four years, making personal investments in my patients, joined for this period of time in our mutual efforts to withstand their diseases, sooner or later it would happen, the disease would take control.

Disheartened, I would have to reckon with my emotions, convincing myself that all the compassion I had felt during the countless visits, all the efforts I had made to develop a trusting relationship with these brave people and their concerned family members, all the time and energy that I had spent diagnosing, treating, trying to control the disease, all this —-all my work—-was not futile.

Dealing with such constant impotence against a disease that resists treatment is terribly sad and frustrating. Not only do I share in the tragedy of the family's losing a loved one, but I also feel like a failure. It tears me up and makes me angry. It's always hard to make this transition to accept the dying process.

I had hoped I wouldn't have to make this transition with Lynn. Now I was afraid the possibility of failure was real.

A week later, after her bone marrow was harvested and cryopreserved, Lynn underwent pericardial surgery. Immediately slides were made and reviewed by pathology. Results were expected within a few days. All this time I was fearful that the worst had happened, that her aggressive lymphoma had indeed recurred. This meant we would have to take more drastic measures to fight the disease. The bone marrow transplantation seemed to be inevitable now...if we had enough time.

When the results were back, Ben called me. "Lynn is recovering well, and the surgeon assures us that he saw no tumor." His monotone voice was putting me more on edge. I was afraid he had raised his emotional shield as a guard against the disappointment he was about to announce. This scared me. Instead, the information he reported was good. "Pathology seems to concur with his observations. All specimens are negative. Whatever was causing her symptoms was not disease."

"Thank God! Then we haven't lost any ground, yet." I exhaled in relief. I certainly couldn't hide my personal gratification. "Do they know what caused the pains?"

"Cardiology is now suggesting that she could have been experiencing these as side effects from weaning off the prednisone." Ben responded dryly. "That's probably going to be the only explanation we'll have for her symptoms." His voice was lifeless, his emotion completely drained out.

"Well, I'm encouraged with this development," I replied, wondering what was on Ben's mind to make him so sullen. It couldn't be this case. Aware that the dichotomies of emotions were often extreme in our practice, I still had to question him about his mood. "What's wrong?" I said with concern, "You sound unusually sad."

"Sorry it shows. It has nothing to do with the Moline case. You know what it's like when you have one train wreck after another? It's been a terrible day," he admitted with disgust and then changed the subject back to Lynn. "We'll give her some time to recover from the chest surgery," Ben remarked flatly, obviously trying to avoid the previous topic, "by that time randomization should be complete and we can start her on consolidation. Herman Gunan will let you know what the computer has selected in a few days."

"I appreciate your help, Ben." I commiserated, granting him his wish for privacy.

"I know you do," he answered. "It's because you do, that I can give you a little more."

* * * *

Days later, Gunan called to report on the computer's selection. Lynn had been randomized for the L-17M protocol, not the bone marrow transplantation. "Now it's in your hands." Gunan seemed glad to bequeath the full responsibility of Lynn's treatment over to me.

For me, the hardest decisions had been made. At least, for now, I would pursue the decided course by trusting the L-17M to do its work, and hope.

By the time Lynn recovered from the two operations at Memorial, she had been off chemo for a total of nine weeks. I was edgy, but optimistic. If her aggressive lymphoma was going to recur, my gut feelings were that it would have done so during her prolonged hiatus from treatment. Since it had stayed in remission—there was no evidence of disease anywhere—our chances were greater that her lymphoma would stay quiescent. However, I wasn't about to take any more risks; it was imperative we start her immediately on the L-l7M consolidation.

For Lynn, the hardest challenge was about to begin. The induction stage of her treatment plan was not easy, but consolidation proved even more difficult for her. Admitted to the hospital on Memorial Day Weekend, she commenced the second stage of treatment.

It was intense. Chemo dripped regularly for five days into her veins, while she suffered through the side effects. At first, she tended to act congenial even to visitors who looked in on her roommate. Usually attempting to appear presentable, she would arranged her kerchief as fashionably as possible over her thinning hair, and despite the tubes which connected her arms to the bags suspended above her on poles, she maintained her good cheer. Her optimistic attitude coupled with a strong conviction in her purpose for living, helped Lynn endure.

Gradually the chemo wrought more drastic changes in her appearance. Her pale face became more noticeably haggard, dark circles circumscribed her sinking eyes, all residual hair fell out, including her eyelashes and eyebrows. Yet despite these frightening alterations, her husband, family and friends, fortified her courage with their love and support. In return, she tried to play down the nausea and discomfort by clowning or joking about what she felt was her resemblance to a bald and featureless store mannequin. Most of all, because she believed in success—the treatment happened to be the necessary means to that rewarding end of her complete recovery—-she tried to keep the spirits of those around her high.

However, the days dragged on. Finally the toxicity of the drugs ebbed through every vessel of her blood stream. She couldn't read, she couldn't relax, yet she didn't have the energy to get up and walk about. Eventually bored by the monotony of her discomfort, she found herself dozing the time away.

By the end of her five-day infusion, she deliberately hid from the all-consuming side effects, sleeping constantly. She paid no attention to the guests of her roommate and no longer feigned good health or humor for the sake of strangers. Although for family and friends, she still mustered her energy to greet them, she was often too tired to care, too tired to keep her eyes open, and even too tired to vomit.

In early June, Lynn finished the first installment of consolidation and was discharged. The chemo had done its work. She left the hospital in very weakened condition.

At home, it was expected that Lynn would recover from the infusion. Instead, she got worse. Within four short days her condition deteriorated. Her dangerously low blood counts plummeted, making her anemic as well as extremely susceptible to infections from common germs. Excruciating sores, side effects of the treatment, coated her throat and mouth and her tongue became covered with rapidly growing fungi. Swallowing became an impossibility. Food and drink were unthinkable. Speech was reduced to guttural sounds; her tongue felt too thick to manage articulation.

If her condition persisted much longer, she would dehydrate rapidly. Although it was my day off when her anxious family called our office, my associate correctly advised them to take immediate action. Lynn was rushed to emergency and admitted.

We treated Lynn with bedrest, strong broad-spectrum intravenous antibiotics, and a special preparation to both soothe and numb her mouth. She also required multiple blood transfusions as she became so severely anemic that, without them, not enough oxygen would reach her tissues.

Even in the hospital under expert supervision, attached to IVs for antibiotics, fluids, and nourishment, her recovery seemed imperceptible. She remained listless and feeble; the stomatitis in her mouth was agony. Overpowering fatigue made her too weak to care and too weary to cry. When she woke from her frequent and deep slumbers, Lynn could only use facial expressions or hand gestures to communicate her immediate needs to her family or the nurses around her. Her distraught family questioned each of my associates who made rounds, "Is she dying?"

"No," they would respond sympathetically, offering clinical explanations as consolation. It was true. She wasn't dying yet, but she was in serious condition, serious enough to make her family and friends think she was dying.

Three days after she was admitted, I walked into Lynn's room and could hardly believe the figure in the bed. It was Lynn, but not the ambulatory usual, I-can-handle-whatever-you-have-to-throw-at-me Lynn. This was a dejected, glassy-eyed, lethargic woman who seemed to be hanging on by a thread. Many questions went through my head at once. Would she die from this episode? Possibly, but unlikely as the antibiotics seemed to have her fever under good control. Would she have enough stamina mentally and physically to continue on with her treatment? I still had faith in Lynn's fortitude and determination even if it needed to be bolstered by my own.

Seating myself at her bedside, I sighed sympathetically and then spoke with a soothing voice. "It may be hard to believe right now, but this will pass, Lynn." I comforted her kindly, distressed to see her rendered this low by treatment. "I'm sorry you have to go through this torment." I checked the labels on the IV bottles. "These antibiotics," I continued, "are warding off infections and it will take just a little longer before the medication will help relieve the sores in your mouth."

She wearily grunted her acknowledgment. Then she formed words which sounded like, "How long?" I understood her question to mean how long would she have to be in this condition.

It was pitiful. I couldn't answer her immediately. Instead, I bowed my head to check her chart. From what I could read about her condition, her descending blood counts had slowed, but it was still uncertain whether they had stabilized or how long she would remain like this.

"Your immune system is way down right now," I answered truthfully with obvious concern; I knew she wanted the explanation for why she was so sick; knowing the reasons for things had always helped her cope. "But, it's not completely wiped out. It will take some time, at least a week before you'll notice any improvement or start to feel better. I'm sorry it won't be sooner." I patted her shoulder gently to encourage her. She had been so brave up to this point, I wanted her to take heart and fight her way back to health. I also wanted her to trust me, to know that I would do everything in my power to care for her. Continuing softly I reminded her, "Hang in there, you're young and strong, and you're going to be okay. Give your body time, time to start replenishing itself." The pathetic figure in the bed nodded that she believed me. Gratified that I had at least given her some emotional comfort despite her physical pain, I mumbled a quiet good-bye, before I left her room.

As painful as it was to watch her suffer, I felt my fury mounting. I was angry over the tremendous disappointment I was feeling, fearful about the possibility of failure. This shouldn't be happening, I thought to myself while heading to the nurses' station in the ward. Had something gone wrong with the treatment? Why was she hit so hard? Perhaps her recuperation from the two operations she had in May was slower than we had realized. Maybe she wasn't physically ready to begin consolidation. Whatever the reasons, I was prepared to take every conceivable action to bring her back.

At the nurses' station, I quickly ordered reverse isolation for Lynn and started her on a stronger battery of antibiotics. Taking every precaution, I called in an infectious-disease specialist to monitor her condition, keeping my own fingers crossed that I had gambled correctly about the treatment and her ability to withstand the worst of the side effects. At least until now, all evidence of disease had disappeared. Since her physical recovery depended on her body's resiliency, what she needed most from me at this time was solace and comfort emotionally. With the assistance of alert hospital care, we could only hope the side effects wouldn't get worse. It would be especially tragic to loose her to chemo.

For ten days, she remained under close surveillance by the hospital staff, examined regularly by me or one of my associates who made rounds, and had around-the-clock support from her vigilant family and friends. To Lynn, each day seemed like the day before—as if there were no tangible progress.

Until the morning Lynn scribbled a note on a pad and handed it to her mother who had been sitting near her bedside. "I'm sick and tired of being sick and tired!" she wrote. Lynn's face expressed her misery.

MaryRae understood Lynn. "It's alright. You can cry."

With enough strength to indulge in a moment of self-pity, anger, and frustration, Lynn regained her fighting spirit. From that point on, she made rapid improvement.

Two days later, Lynn's improvement was matched by the exponential rise of her white cell count. Once it was normal and she regained her strength, we were able to discontinue her intravenous antibiotics. Her mouth sores had almost completely disappeared and she was able to begin eating pureed foods. A day later, she actually chewed her first real meal in over fifteen days. With her renewed vitality her sense of humor also returned. By the first full day of summer, she was able to be discharged. Everyone was delighted, except me.

I was worried. I couldn't allow myself the privilege of sharing everyone's relief at her recovery from this crisis because I was already concerned about the next hurdle. She had barely weathered through this phase. There still was a second infusion to come. Would she be strong enough to take it?

I reflected on the field of medicine I had chosen. It causes such suffering and severe toxicity in order to effect its benefits. I felt remorse that it often had to be this way for my patients, Yet, I had no guilt that I was in any way responsible for this toxicity. I always warned my patients that problems were possible and sometimes even inevitable. But, I longed for the day when we would have drugs that would be more specific for the tumor cells that they were designed to kill, thereby limiting the adverse effects on the normal cells. This and the promise for more effective drugs were part of the excitement that led me into this field. For me there was no greater thrill in medicine than watching a malignancy disappear as a result of the careful and planned application of my craft.

Lynn's case had to be handled precisely right for it to be such the success I hoped it would be. Even though I trusted the L-17M protocol, silently I concluded that the dose may have been too strong for her. I would review the protocol and check the doses. Maybe the amount of drugs could be reduced to prevent her from enduring this particular torture again. Maybe... there were no guarantees.

* * * *

"I think your ready again." I told Lynn one morning when she had returned for a routine examination and CBC (Complete Blood Count). As I leaned against the counter top to write my observations down in her chart, I explained. "Your blood counts are now way up and everything else looks good. You're due for your second infusion." It had been almost a full month since Lynn was discharged from the hospital. She hadn't had any treatment during this period of recovery.

Hopping off the examining table, she reacted to my words with obvious disdain. "Ugh," she grimaced. "Can't say I'm looking forward to going through it again."

I understood Lynn very well now. Unlike some of my patients who always needed tremendous coaxing and pleading to get them through treatment, Lynn was expressing an honest reaction, not fighting the decision. I knew Lynn had confidence in the effectiveness of the chemotherapy regimen and had accepted it as vitally necessary for her survival.

"You may not have to," I assured her, "I'm reducing the ratio of dose to body mass so as to retain the potency without the drastic side effects you experienced the first time around. You were really zapped, and in the long run, as bad as it was, it may turn out to be a good thing. The harder we hit aggressive lymphomas on the onset, the greater the chance for cure. Undoubtedly you were hit hard, personally I feel it was hard enough to allow us to reduce the formula so you won't have to go through that same ordeal."

"That's encouraging," Lynn quipped and then asked for details on drugs and doses according to the revised plan. Since her ordeal in June, Lynn wanted to be as knowledgeable as possible for her own protection.

Happy to oblige her with as much as she could absorb, I afterwards recommended immediate admission. "Tomorrow, so we can get this part over with."

Reluctant as she was to move so swiftly, Lynn conceded, "The sooner I get this over with, the better!"

It was for the best. Although we feared that even the reduced amounts would affect her as before, our fears were unfounded. Lynn weathered the second phase of consolidation with less toxicity and her recovery was quicker. The decision to change the formula had been right.

By August, the hardest challenges of Lynn's journey toward cure had ended. All disease had vanished, all side effects had waned, and all intense treatment was officially over.

* * * *

"I have a bombshell to drop." Lynn said shyly one day during the routine office visit. I could read by the expression on her healthy face that she wasn't about to deliver bad news. Instead she more resembled the proverbial cat that had swallowed the canary.

Two full years had now passed since she had first been diagnosed and started chemo. After consolidation had ended in the summer of that first year, prophylactic radiation followed as a course of treatment. Then, by that fall, Lynn went on maintenance chemotherapy—the final phase of her protocol—for almost a full year, juggling both normal life and work with the occasional disruptions from the maintenance regimen.

In September of the following year, we discussed together whether she should continue with maintenance treatments. There was sufficient reason for discontinuing therapy, in my opinion. Lynn suffered from chronic low blood counts which prevented me from administering the full dose recommended in maintenance. I began to feel it was almost useless to continue treating her at such low doses. In addition, recent findings indicated that prolonged maintenance for aggressive lymphomas was not as essential in achieving cure as it was once believed. After weighing the pros and cons, Lynn agreed.

Now, it was five months since Lynn had been off any kind of therapy. Although during these two years, her cancer never once recurred, I was still fearful that if she relapsed, it would happen in this first year off treatment. So far, she had remained in complete remission.

"What's this bombshell?" I asked with calm curiosity while seating myself on a stool in the examining room, flipping through the thick file that contained all records of her case.

Lynn hesitated, noticeably embarrassed, before she blurted out, "I'm pregnant!"

I was taken by surprise. Speechless, I could only laugh.

"I'm glad you're laughing," she responded with relief, "I thought you'd be angry."

Actually I was merely astonished. Aware that during treatment, Lynn's menstrual cycle had become erratic, I had supposed that she was in the age group for which sterility would be a strong possibility.

Depending on the type of cancer, it was unwise for a patient in remission to become pregnant. Some cancers could be stimulated to recur by the hormone activity of the pregnancy. Other cancers, such as the one Lynn had, would not be affected by the pregnancy itself, but if relapse occurred while the patient was pregnant, there would be moral and ethical questions to consider about how chemo would affect the developing fetus if treatment was resumed. It would make a terrible situation, horrible.

During the two years of treatment, Lynn had constantly inquired and probed about the potentials of childbirth, so she was well aware of all these arguments about pregnancy. Still she admitted, "It's a miracle! I sincerely believe it." She fumbled for words, "You know we had wanted children before I got sick, but we were intending to wait until I was considered cured—which would be in almost three years—as you recommended, and since I'm already thirty-two years old, I would have been thirty-five by the time I could start my family. Truthfully, I couldn't be happier with this surprise... I guess God had other plans." she shrugged.

"Are you sure you're pregnant?" I asked hoarsely and cleared my throat.

Lynn nodded emphatically, her short brown hair danced on her bobbing head, "I took an early pregnancy test, twice!"

This was an amazing development. "Well, congratulations! I'm happy for you too." I said sincerely once I found my voice. "We'll do the blood work here to make it official," I added. Resigned to accept whatever the future held, I didn't want to surrender to the strong misgivings that haunted me. "Since you're aware of the medical concerns, we might as well take this as a good sign of your returned health and leave it at that. I can recommend a good ob-gyn group which has helped several of my patients, if you're interested." At that moment, I couldn't bear the thought of tragedy.

"That'd be wonderful!" Lynn replied appreciatively.

Even though I had become cautiously comfortable about her chances of survival, I still silently worried about recurrence whenever I examined her. If I were to discover a big node sticking out and she was to admit, `I just found this,' I would have had to walk out of the room——I'd be totally devastated. It would mean I was a failure (even though I believed Lynn had the best medical care available). It would also mean that because this poor young woman had relapsed, the chances were greater that she would die from her disease. Then, I would have to be a part of watching her slowly succumb, I would have to take her through to the end as I have with so many others.

It would hurt tremendously because our mutual affection and ties had grown stronger with time. Sometimes I have cried for my patients at such a turn of events and I think many of them are touched by this genuine flow of emotion. Knowing that I share their concerns helps them deal with the difficulties that lie ahead—there is no substitute for this bond of true caring for a patient by a physician.

However, for other patients who look upon me as some kind of emotional Rock of Gibraltar, my tears would only serve to frighten them. They might even question my professional ability to treat them.

Now, Lynn's pregnancy complicated things even more; it made the possibility of tragedy even more intolerable.

"How are you feeling?" I asked hesitantly. My sincere delight in her good fortune was unavoidably mixed with great trepidation.

"Don't worry," Lynn assured me, "I'm really fine. Sometimes, I feel like a tiger ready to pounce on everything and everybody. It's probably the crazy hormone changes. Until this happened, I was merely feeling invincible." she smiled.

Fortunately, as it turned out, Lynn was still in complete remission. After I completed her examination, I relaxed and asked, "How far along in your pregnancy are you?"

"I guess about two months." Then her smile broadened. "It's coincidental that I'm due in September which will be the first anniversary of my remission, because just in this past month, before I realized I was pregnant, I started worrying so much about making it to September without recurrence, that I was very frightened. The more fear I felt the more afraid I was that fear itself would bring on a relapse—among the other things that have helped me along, I believed the power of my mind played an important part in my cure—so, I didn't like being paranoid. Now my outlook has definitely changed. Now I have an anniversary of ending treatment and a baby to look forward to in September."

I nodded my approval. She always seemed to find a way to turn the situation around and look at the bright side. "Off the record, Lynn." I admitted candidly, "my gut feelings say you're already cured. We hit you pretty hard at the beginning, and that was the best way to treat whatever kind of aggressive lymphoma you had. You're already almost halfway through the first year entirely off chemotherapy and everything's clean. I know you find this encouraging and so do I. It's very encouraging. With regular checkups throughout the next..." I glanced at the chart for the precise figure, "two years and seven months, you'll be completing the three-year remission period soon. After that," I reminded her, "the survival curve flattens—which means you'll be considered cured." My own hopes were high about her chances for success.

The three-year remission period melted away.

In September, having completed the first full year of remission, Lynn gave birth to a happy baby girl.

When I paid her a social visit in maternity a few days after the birth, it was a triumphant moment for the both of us. "So how does it feel," I asked her, "to be back in the hospital under these conditions?"

The joy written all over her face was an answer in itself. "Wonderful!" she smiled, lovingly cradling her newborn daughter. The happy tears of uncontrolled emotions moistened her eyes; it was one of the few times she let her emotional guard down for me. "This moment would not have been possible for Russ and me if it weren't for you!" she declared. "There aren't words enough to thank you for my life and hers!" Overcome with profound gratitude and affection, she shyly looked down at her baby, giving her a hug.

I understood what she was saying and savored the warm sense of gratification I was feeling. This was a reward I always needed.

In September of the following year, Lynn completed the second full year of remission and at the same time announced that she and Russ expected a second baby in May.

Another September later, completing her third and final year in remission, Lynn, now the proud mother of a two-year-old daughter and a four-month-old son, was officially considered cured.

There was no great outward fanfare at the office. Routinely the date of her visit and the remarks about her current health status were faithfully logged on her progress sheet. It was inappropriate to be ostensibly jubilant over success in the presence of so many other sick patients seeking miracles—miracles which as yet don't exist for many of them. Despite the great advances in medicine, we still don't have the drugs or medical technology to effect cure in the majority of our cancer patients.

Although it's unfortunate that cure can't happen more frequently, I am thrilled by the private joy of knowing that there are healthy, happy people walking around out there because of my medical expertise. People I have grown attached to and care about are now fulfilling their hopes and dreams, which they would not have been able to enjoy without my professional success. It's always a great pleasure to be part of that special process of cure. It's one of my greatest joys. It rekindles my enthusiasm and devotion, making me feel as though my life's work has meaning and significance. Most of all, it reminds me that, when the conditions are right, cancer can be vanquished completely. That is why I have such a warm sense of accomplishment and pleasure when I reflect on these curable cases or see these former patients picking up where they left off before the disease invaded their lives.

Lynn and Russ have indeed picked up with their lives. Eternally grateful beyond words for this opportunity to begin again, Lynn, with her husband and now three young children, remains in good health.

For some, life goes on.

4~

The Serenity To Accept

Bureaucratic interference imposed on my practice by government agencies and health insurance companies made some decisions more complicated. Whenever I've had to make a choice between treating a patient or letting go, I resented the added pressures of restrictions that seemed unrealistic, impractical, and counterproductive. Rather than letting a patient die with dignity, often I found that the patient would be victimized by the legalities of dying.

One morning on rounds at the hospitals, I was to face two such difficult cases. It was already 8:10 A.M. on a cold December day and I was running late. Trying to locate a patient who had been moved was putting me behind schedule and making me anxious.

"Oh... We transferred Mrs. Miller to surgical recovery, Doctor Berger." The nurse replied quickly to my question as she hurried down the curved corridor of the University Hospital. I could see that a flashing light over a far room in the ward was her destination and let her continue without questioning why. Although, I had to get back to my office patients by 10:00 A.M., it was important that I check on this patient wherever she was.

On my way to surgical recovery, I mentally reviewed Mrs. Miller's history. She had been with our group for treatment of an aggressive metastatic breast cancer. Because she was very elderly, she had elected to discontinue further chemotherapy when it was determined that the risks of toxic side effects were no longer worth the benefits or the chances of improvement. It was only a matter of time before the cancer would soon kill her, but no one in our group could really disagree with her decision. She had virtually no family left except for a niece who would drive her on occasion to our office for visits. Otherwise, she lived alone. Respecting her wishes, we ended treatment.

However, more recently, she had developed bleeding esophageal varices (varicose veins) for which she needed an operation and had been admitted to the University Hospital under a gastroenterologist's care. Apparently, this operation to sclerose or coagulate her varices was somewhat of a research procedure at that time. Our group was ever watchful of her situation, but as we were no longer her primary physicians during this particular development, we had to keep in the background for the most part. This was professional courtesy.

When I arrived in the unit, I noticed a crowd of doctors, all in lab coats, encircling her bed. Some were nodding, others talking, while the patient lay motionless. Tubes, bottles, and bags seemed attached everywhere. Although the old woman was not alone in the unit, she was the only patient at the center of such attention. My impression, as I approached them, was unfavorable. I had seen this scenario before and it disturbed me.

"What's the problem? " I asked politely after exchanging greetings.

"Well, this patient is quite anemic after hemorrhaging and will be needing two units of packed cells..." the hematology fellow offered in answer to my question. She then explained what was happening. Although the first attempt had been partially successful, it became necessary to clot off the veins that were bleeding in her esophagus a second time. A condition of severe cirrhosis, from which the patient chronically suffered, had left her with very few clotting factors, and postoperatively, she had begun bleeding heavily.

As I listened, I remained silent. It was obvious the kind of care which they outlined for the patient was planned with every good intention. Certainly, it afforded these young physicians-in-training the opportunity to gain hands-on experience with an experimental technique, but I was failing to see how its repeated application could truly help Mrs. Miller in the long run.

The poor old woman was already dying of a progressive and malignant cancer, and now it looked as though, despite the operation to stem the flow, she was bleeding to death from the varices themselves. How many times would her physicians be willing to subject her to this procedure just to keep her alive so she could die of her painful cancer? Spending all this time, effort, and expense on a patient whose recovery, in my opinion, was so limited seemed very futile to me. As I had no alternate therapy to treat the cancer which was killing her, and as she had already accepted death as a welcomed alternative to lingering in extensive pain, I personally could not find justification in having her life prolonged through a series of experimental procedures.

In a case like Mrs. Miller's, the line between life and death was not clearly defined. She was already dying when these doctors brought her into the operating room, and she was still dying when they brought her out. They had not changed that fact. Her advanced age was also a detail to consider. Recovery from repeated surgery for someone as old as she and in a pre-terminal condition from end-stage cancer was improbable.

Which was the better way to die? Given Mrs. Miller's actual choices, bleeding to death, in my opinion, was quicker, certainly painless, and therefore more humane.

I had to ask myself: Were her doctors at this point, interceding to her detriment? When was it torture and when was it cure? Should these life-saving measures be applied, or were there legitimate times when withholding extraordinary procedures was an ethical alternative? Did she really want them to put her through so many more operations?

Apparently, her doctors could only see the short-term result of the procedure, that she was bleeding severely, and so they were doing their best to stave off death. It was hard for them to consider the long-term prognosis of her cancer as a factor in their treatment.

Perhaps to them, it was an open-and-shut textbook case, but to me, it was more a judgment call based on actual experience in practice. Despite the clinical arguments for the repeated procedure, the real question they should have been asking was: What did this patient really want?

Apparently her doctors didn't know. But I did. I had come to know her well during our years together fighting off her disease and I felt I knew her wishes. As long as Mrs. Miller had no family to represent her, I felt compelled to speak up. "Well, I think you're planning to go too far with Mrs. Miller," I began, despite the fact that I was only a consultant and not expected to have an active hand in her care. "The poor lady can't speak for herself right now, but she already decided against prolonging her life needlessly when she discontinued her chemotherapy treatment with us. Now, here she lies with tubes in every orifice, when she was hoping for a more simple and quick death all along. She may have given you her permission to sclerose her varices when she signed that consent form, but knowing her as I do, I don't believe she wanted you to go all out for her given the current circumstances."

They were all listening quietly.

"When our group was treating Mrs. Miller for her breast cancer," I continued, "she expressed for herself that she didn't want her death complicated by extraordinary means. At that time, I had promised her we would provide her with symptomatic and supportive care without unwarranted heroics. That's why we ended treatment. She wanted it that way."

I paused to check their reactions and saw all serious faces. I didn't enjoy holding the minority opinion. "Understandably, the situation is a bit different now with bleeding varices, but remember that the bottom line is that she still has a very progressive and terminal cancer. Her death from this cancer will be painful... more painful than if she should bleed out. Keep in mind, whatever you do, the main question which she has already answered for us: What has she to live for?"

My aim was to persuade them to think of other possibilities —that there were indeed other options to explore rather than to transfuse the dying old woman relentlessly—although I realized that not everybody tended to think like I did.

Sometimes for young or new doctors, such as this group, the larger picture was not so visible. They were more focused on the act of saving lives and the sense of triumph and reassurance that success would bring. Allowing death to overcome a patient would usually be considered the ultimate failure.

"We appreciate this information," the hematology fellow replied coolly after it became obvious no one else was going to speak. The others grunted their confirmation almost in unison. "As you know," she continued, "we will do our best to make her comfortable in whatever way we can. Our intention is to improve her condition at least by stabilizing her. From there on, it's out of our hands." I translated this to mean that the input I had offered would be analyzed, the conclusion would still have to be determined.

I expected this reaction. Nobody likes being affronted or criticized, and I had a habit of being blunt, but I felt it was necessary to put the whole situation into perspective. "At least get the patient to a lucid state, then ask her what she wants you to do." My only advantage was that, from their point of view, I was a physician with seniority. That gave me authority along with the fact that I had prior knowledge of the patient. "Certainly no one can fault you then." I reminded them.

"Thank you, Doctor," she replied curtly and then directed the group over to another patient.

Mentally, I gave a shrug. I could do no more. I didn't have the energy to take on the world. They were entitled to their opinions, although I believed in mine. I felt my stand was ethical. I hoped they understood me and my viewpoint. If not at this stage of their careers, they might eventually understand.

Having expressed my opinion, I felt relieved. I had done all I could as an outside consultant. It was still up to them to decide her fate. I would be having enough emotionally draining decisions waiting for me in the day ahead. I could spare no further time nor energy for poor Mrs. Miller, and for this I felt remorseful.

Flipping through the copious notes in her chart as the medical team moved on, I studied once again the records of her overall condition. It was as I thought, very grave. I closed the binder and leaned over Mrs. Miller's bedside. Unconscious, the old woman seemed almost comatose. I patted her hand and then felt her pulse. It was barely perceptible. Perhaps it was better, this way, for her sake. God willing, she might not survive the procedures in store for her. As awful at this might sound, I had no guilt about my feelings. If she were lucky, time would be mercifully short for her and solve the problem for her. If she were lucky...

* * * *

Feeling troubled about the way Mrs. Miller's case had developed, I left the University in sour spirits, only to find that rounds at the next hospital were even worse.

Reading the words scribbled on the progress sheet of a new patient was a shock. I had just collected some of my patients' charts, preparing myself to visit them in their rooms, when I noticed an awful development in one patient's chart. Painfully, I looked up at the bright lights and Christmas decorations dangling over the nurses' station with disbelief written all over my face. "This is awful!" I moaned aloud to someone standing nearby and then looked down again at the chart.

"What is it?" a soft voice replied to my distress.

"Who's in charge of this patient, Dominic Vincente?" I asked in general, hoping one of the nurses nearby might clear up my momentary bewilderment. Perhaps I was confusing him with a different lymphoma patient at the end-stage of the disease.

"Oh, I am." The same soft voice replied. I looked over at a new nurse who stepped forward. Quickly I checked her badge for her name.

"Do you know his history, Karen?" I asked. "I know a Vincente who was only diagnosed for early-stage of his disease four weeks ago. Is this the same patient? A man in his early sixties who has just started his chemotherapy at Memorial for a very bad T-cell lymphoma?"

"Yes," she nodded without hesitation, "He had been discharged from Memorial early last week and was admitted here about four days ago. I was just going over his chart."

It was not unusual to have patients transferred back and forth in this manner. Our group has sent patients to Memorial Sloan-Kettering to determine the most advanced treatment available for the particular disease, and then we receive the patients back providing them with the same treatments with the added convenience of being near their homes. This has generally been a very successful collaboration.

What was unusual was that the man's condition had grossly deteriorated in such little time. His white cell count was very low; he was also jaundiced and had renal failure. It was becoming apparent that his generally poor condition from the lymphoma had left him with little reserve to fight the toxic side effects of the chemotherapy. The only good piece of information written in the progress sheet was that the cancerous nodes had cleared, the tumors had disappeared.

Karen continued. "It's all been pretty fast. Doctor Chandler and Doctor Smith have been looking in on him, along with your partners." She shook her head. "They feel it's grave."

"I can see that," I answered with obvious frustration while reviewing the comments of the nephrologist, the infectious disease specialist, and my partners.

"What can be done for him now?" she questioned sadly.

"Nothing more than what's already being done: IV fluids, hyperalimentation, broad spectrum antibiotics to cover a wide variety of infections, and careful monitoring of his electrolytes." I recited what had been recommended by my colleagues. "But, truthfully, it doesn't look good."

Room 304, the chart was labeled. I felt shaken and reluctant to see him, feeling partially responsible for this awful outcome. As I proceeded toward his room, I had to question myself whether the risks had been worth it and whether the toxicity could be reversed. Would there be any chance of a viable recovery for this patient? Because the disease had so quickly disappeared, the potential for remission was the only tangible clue at the moment that we had taken the proper course by administering the chemotherapy to him. Yet, despite that glimmer of hope, the fact that he had suffered such severe adverse effects held an element of surprise.

Drawing from my years of experience, I knew that when a patient's initial response to chemo was this drastic, the overall prognosis was generally poor. Making the "right" choice, always extracted an emotional and psychological toll. Sometimes I feared becoming a therapeutic nihilist. Even so, there would have to be more seemingly irreversible complications before I would sanction with a clear conscience the patient's right to die, that is, hold off on chemo.

At the time of his initial consult with me in my office only a few weeks ago, we had discussed the problems with chemotherapy and the major risks. "... hemorrhage, major organ failure, infections, respiratory failure..., " I said matter-of-factly giving him the routine rundown. "The list continues with some lesser effects, such as hair loss, nausea, constipation, lowered blood counts, cystitis..." It was a common litany, repeated frequently through the day. While listing these side effects, I never suspected such sudden tragedy for him.

Yet, he was very fearful. I distinctly remember how immobile he was as he sat next to his wife across the desk from me. He had a shiny bald head and smooth round face, which probably at one time, gave a jolly appearance. However, that was no longer the case. This man was sick and frightened and very sad, and I could tell by his posture and listlessness that the disease was ravaging him quickly.

I understood his fear. Hearing the dangers from the toxicity of chemo sounded almost as bad as the disease itself, but as with all my other patients, I quickly reassured him, "I know all this sounds shocking. Generally the degree and severity of toxicity vary from person to person. Some of my patients hardly feel the side effects, a few may have an extremely bad time, but most manage to get through it all. Keep in mind that the potential for cure outweighs these risks." This was true most of the time. Regrettably now it was not the truth for this man.

The memory of those words, as I entered his hospital room, made me dread seeing him in his present condition. His sad wife was sitting by his bedside and stood automatically when she saw me. "Hello, Mrs. Vincente." I said somewhat solemnly, as I went over to his bedside. He was asleep, although he was so still, it was if he were unconscious. Realizing from the way things appeared that his chances for recovery were poor and that he was possibly near death, I was mentally preparing myself to explain the unfortunate circumstances to his wife.

First I examined him to confirm what was written in his chart for the past few days. His blood pressure was a very low 70 over 50. He was jaundiced and only marginally lucid. Now, even after several days of antibiotics, he was not responding. His infection was not under control. I checked and double-checked the progress sheets, flipping back and forth through the pages as I sat quietly beside the sleeping man. I was too troubled to speak yet to his wife.

Sick with lymphoma before treatments began, Mr. Vincente had become worse from the side effects of the chemo. As an older man, he had less energy to fight off either his disease or the toxicity of the medicines. There was no chance for heroics here. Had he been completely lucid, I might have encouraged him to fight. But I knew instead, the way things were going, he would be chronically ill from the disease and incapable of effectively combating it.

The decision I was about to make for him was one of the hardest I ever have to make for any of my patients. If his family agreed to it, I was going to recommend that he be coded DNR, (Do Not Resuscitate), which would allow him to die without extraordinary medical intervention should he suffer respiratory or cardiac arrest.

I felt compelled by the information at hand to make this life-and-death decision, in other words, to play God. This was a burden of responsibility I dreaded, but my concern was for the comfort of Mr. Vincente. As much as I was not making this decision in a vacuum (I had the concurring documentation of my colleagues literally at my fingertips), I realized it had to be taken one step further, it had to be taken out of limbo. That decision had to be reached soon, because if we waited much longer the patient would be progressively subject to undue suffering. Before I could come to such a conclusion, I had to be convinced that this decision would be the right one. This was to be determined by input from Mrs. Vincente.

Looking up at Mrs. Vincente, I kept my face impassive. "Let's step outside," I said calmly and gestured toward the corridor. I followed the diminutive woman out of the room as I prepared myself for our conversation.

It is no coincidence that the potential for long-term recovery or cure from cancer resides mostly with the young. Among the young patients I have treated, many have had a decent chance of turning the cancer around with the proper treatment. That is when I do everything possible to convince them that I can bring them back to health. Even if they lack a positive attitude, success in treatment is more likely for them. That is because, generally, young bodies can physically tolerate these extraordinary attempts we may be forced to undertake. And frequent success is often our reward.

Yet, the hazards have always been greater for my older, severely diseased patients. Generally for them, the curative rewards from chemo do not outweigh the risks. Because some of them do not have the drive they need to endure treatments or else lack the physical, emotional, and psychological strength to deal with the adversity which lies ahead, I find I am less inclined to put the unwilling ones to the test. Combined with facing the fact that an older body cannot endure as much punishment, I often find I tend not to struggle as hard, not to put these patients to task, not to push beyond the limits of their willingness, age, and human endurance for any heroic attempts toward a difficult cure. That has been when I feel most vulnerable. That has been when I feel most culpable and most responsible. Have I tended to be too willing to let them die?

At such times, I try to put myself in the patient's shoes, asking myself, "What would I want when I am old and dying of a terminal disease?" Sometimes, I see that hospital bed as my own inevitable destiny and it frightens me to think that I could be in such hell from the endless and unendurable pain, with a foot in the grave, waiting to die, and some idiot decides to pump intravenous antibiotics into me in order to stave off a potentially harmful infection—to keep me alive further when that infection could have mercifully ended it all for me sooner. Just thinking about that would make me livid!

While my greatest hope has been that I have been right in allowing a patient to die, so my greatest fear has been that the tenacity of some people, whether old or young, to hold onto life regardless of its quality might be greater than I had ever suspected. Then the tragedy of their deaths would be unspeakable.

Who truly has the right to decide who should live and who should die based on criteria of what is curable and what is not? This is when the doctor's role over the life of terminally ill patients may be godlike. Yet, for physicians, it is an awful, painful dilemma, and more times an undesirable responsibility to bear.

Even though I have felt that, regardless of the age factor, the patient—not the doctor—has the ultimate right in deciding his or her future, I do believe that, sometimes, doctors are in the better position of knowing the future suffering that these patients will endure. If through their experience, the worst complications of the disease can be avoided, and if the patient seems ready to accept death as a desirable alternative to greater torment, then it is the tremendous responsibility of the physician to guide the dying to an easier end.

Still, if the patient were informed of his or her alternatives and had the courage to suffer all the consequences in the hope of even marginal success, then the doctor's responsibility is one of assistance. The fate of the patient remains totally and unquestionably in the patient's own hands.

While my mind had been leaping through this obstacle course of ethical and moral hurdles, the actual pause was hardly perceptible to Mrs. Vincente. As we reached a location in the hall not too distant from her husband's room, I began with an apology, "I'm sorry, Mrs. Vincente," it was a somber moment. "Your husband is not responding well to the antibiotics, his condition is critical, and his chances of recovery are unfortunately very poor."

She nodded silently as her moist brown eyes stared up at me. I realized I was towering over her by more than a foot. "Let's go over there to the visitor's room to talk." I offered, figuring we could sit down in the chairs and considerably level off the height disparity.

"What happens now?" she questioned quietly when we sat down. Her small face seemed more wrinkled and gray as if she had aged ten years since our first meeting a few weeks ago.

"If it comes to a crisis," I answered slowly, "we can keep him alive if you wish by life-sustaining technology. As it stands, his survival of the lymphoma is contingent on his ability to tolerate the chemotherapy regimen he has just begun, and obviously he is not tolerating it at all. Any potential to cure the lymphoma is negligible. Even though the tumors have disappeared right now, if he does not continue treatment, he will relapse. As I've explained to you both before, the additional side effects of chemo on kidneys, liver, heart, and other organs are detrimental. In this case, his kidneys and liver are showing signs of deterioration, and even though he is on antibiotics, he is not successfully fighting off infection." I looked sympathetically at this small woman, whose husband of over forty years was being taken from her. "Do you understand everything I have been saying?"

"Yes." she hesitated "I only want to do what is right for him. What should I do? He is not good with suffering. He complains when he gets a stomachache," she was talking out loud to herself, trying to figure the proper course of action. It was a hard decision to make alone.

Yet, through her self-questioning, she was providing me with information to help reaffirm my unspoken decision. If she were to have said, "He has always been a fighter. He would never give up this easily." I would probably have held back on my DNR suggestion a little longer.

"Do you have any other family to discuss this with?"

"Four weeks ago, when he was first diagnosed, we had talked over the possibilities with our children. They're all grown up with families of their own, but they're good children and have been quite supportive."

"What did you and your husband and your children decide?" I questioned soothingly to get her back on track.

She was speechless for a moment. The answer stuck in her throat.

"It's a tough call," I tried to reassure her as I gently reached for her hand. "I don't want to lead you into anything. However, you wouldn't be wrong to feel there is little more we can do for him. The question is whether you wish to take any action to resuscitate him should he go into respiratory or cardiac arrest or to keep him comfortable but let him go without intervention if his heart or breathing should stop." I could see that while her grief for him was tempered with concern for his comfort, she was relying heavily on me to provide her with a proper course of action. "Allowing him to die in this manner requires an order I would have to give a DNR order meaning Do Not Resuscitate. This does not mean we will stop all treatment. This means that only if his heart or his respiration stops, no one will interfere." I had to emphasize this point. I did not want her to believe we would discontinue any vital IV fluids or painkillers. She appeared to be listening thoroughly although she still was unable to utter a sound.

Even though I had not known Mr. Vincente for very long, I was torn by many factors in his case. The responsibility and burden of guiding his wife toward what I believed, and my colleagues believed, was the right and more humane decision were taking its toll on me. I felt a strong rage growing inside me, a rage from frustration and despair. It should not have happened this way! It should not have ended so quickly. It was a lousy and unexpected turn of luck! Yet, I could see no better future for this patient and so could advise no other recourse. It all depended now on Mrs. Vincente. Would she be able to let him go should his heart or respiration fail?

Finally a feeble voice emerged from the tiny woman. "That's what we all thought," she whispered hoarsely, "Let him go in peace. That's what he would have wanted. I really don't know what else to do for him. Oooh, this means he won't even make it to Christmas! " She bowed her gray head and choked back a soft sob.

"I am sorry. If you need a little more time to come to this decision, please tell me." She shook her head. Patting her hand nervously, I tried to contain my own emotions. Even though I felt she herself genuinely reached this decision, in essence, I had been the determining influence in achieving it. It worried me to have such sway, not because of any legal repercussion—although those do exist—but because this was another life I could not help in any other way except by making death easier.

After conveying my utmost sympathy, I left Mrs. Vincente to deal with her own grief. Experiencing such a catharsis for the second time in the morning left me feeling drained. As I headed back to the nurses' station to tend to the practical matters of paperwork, I forced myself to make a quick emotional recovery. If I were to have appeared as depressed and somber for my next patient as I was actually feeling, I would be bringing someone else's tragedy into that patient's life. This wasn't right. When it came to my emotional presentation, each patient deserved to have a clean slate. Yet after two such exhausting episodes, it might just be a little harder to bring me around to lighter spirits.

Now that the DNR decision had been reached, there were forms to fill out so as to ensure all legal loopholes were covered ——-paperwork and forms that seemed a waste of time. Already behind schedule, with other patients still to see and my office waiting for my arrival in less than an hour, I knew the record still had to be set straight. This would protect me later if suddenly the family should decide to turn around and accuse me of forcing the decision, in a malpractice suit. In addition to all the sorrow and misery surrounding such circumstances, it angered me to have to consider these legal repercussions at a moment like this.

"Doctor Berger." A nurse interrupted my thoughts, "Your patient, Rogers, in 307 is ready to go to x-ray. Did you want to see him before they take him down?"

"Are they taking him right now, Jane?" I asked wearily without looking up from my notes.

"They just called to say they're on their way. Do you need a few more minutes to complete that?" Jane questioned while pulling the Rogers records from the chart rack and placing them near me.

"I'm almost finished. Please don't let them take him 'til I see him." I noticed my watch: 8:57 A.M. I was afraid I was going to be late for the office.

After completing my notes, I closed Mr. Vincente's chart and put it on the nurses' desk where the orders would be picked up. Then, I hurried to see Gerald Rogers. A hospital runner was wheeling a chair behind me as I entered his room. X-ray wasn't going to get the patient before I saw him.

* * * *

Trying to shake off the gloomy pallor of this Tuesday morning was not easy. However, despite the way the day had started, most of the patients I visited on rounds afterwards were showing remarkable response to treatment. This restored some of my confidence in my craft. Seeing their smiles and sharing a laugh with them began to lift my spirits.

Then something wonderful happened. I saw a former patient, Joseph Fry, outside in the hall. He was one of my patients who had been cured.

"Hi, Doctor Berger!" Joe waved as he wheeled an older woman, dressed in street clothes, to the elevators behind me.

"Joe! It's nice to see you looking so well." I replied, elated with his appearance.

Joe was a bald young man who eight years earlier was stricken with a germ cell testicular cancer which made the tumor in his groin grow to the size of a softball. By the time I was called in for consultation, the surgeon had already done a radical orchiectomy. However, Joe's chest x-rays were showing pulmonary nodules. Apparently, his cancer had already passed through the venous system to the lungs. After thoroughly researching his options, I chose a protocol that actually could cure him of the cancer, except that, back then, the regimen was particularly harsh.

Joe responded well to treatment. His pulmonary nodules disappeared and finally, he was in complete remission when he suffered a seizure. The CAT scan showed metastases to the brain. Joe, his family, and I were all very upset with this new development. At that point, Joe was preparing to die. I knew that brain metastases were a very bad prognostic sign because it was believed that chemo could not penetrate the blood-brain barrier. This meant that disease in the brain could only be effectively treated with radiation therapy to the entire brain. Reviewing the world's literature on treatment, I decided we should radiate his brain while continuing his chemotherapy. Joe, however, wanted to stop.

But Joe was only twenty years old! I couldn't let him give up without a fight. And personally I didn't feel it was a certainty that he had to die. Somehow in the face of overwhelming odds, I still felt there was reason to hope.

Here was a case where I had to push therapy very hard, harder than Joe was willing to go. I reminded him that there was indeed potential for cure despite my own doubts. I pushed, joked, and even yelled at him telling him that he couldn't quit, that his life depended on treatment.

Battling his lassitude and deepening depression was almost as bad as fighting his cancer. It took great perseverance on his part and mine, but after much give-and-take, we overcame both his cancer and his apathy.

Three years after all treatment had ended, Joe was considered cured. The only sign of his ordeal was his bald head, a permanent side effect of his strong radiation-chemo protocol.

"I'd like you to meet my aunt, Joy Williams." Joe said, introducing me to the gray-haired woman as he parked her wheelchair in waiting near the elevator doors. She was sporting an ankle cast.

"So this is your famous Doctor Berger, Joey." His aunt Joy said, giving me a gracious smile. "I've heard so much about you and what you did for Joey. It's a pleasure to meet you."

I privately basked in her praise and appreciation as I continued chatting with them.

When the elevator doors opened, Joe said, "Good to see you again, Doctor Berger," while his aunt said good-bye, and then they both disappeared behind the shutting doors.

Seeing a patient who was at one time so close to death, now participating so ordinarily in life was a boost to my ego. It made me happy. By the time, I visited my last patient, I had put aside all my sadness from the morning events and was feeling downright cheerful.

My last patient was Mr. Worth, an older man just recently diagnosed with lung cancer. I did not expect any problems since his condition was stable, he was admitted for routine tests and treatment, and he was actually doing quite well. He was supposed to be discharged today. It should have been a piece of cake!

When I went into his room, I was mildly surprised to see he had some visitors, two middle-aged men and a middle-aged woman. As we exchanged introductions, I learned these were Mr. Worth's children. Since the hospital did not have regular visiting hours at this time, I realized they had probably come to take him home once he was discharged.

I went directly over to the patient's bedside. "Good morning, Mr. Worth!" I boomed loudly so he could hear me and clapped him gently on his frail shoulder. "It's a good day for you today! Your fan club is here, and after I give you a quick check, we can send you home. Isn't that good news?" I shot him a pleasant smile.

"Yeash, Doc," he lisped happily, "I'm feelin' guudth." He sputtered as his toothless smile grinned back at me.

"That's great! Let's have a look at you so we can send you home." While pulling out my stethoscope, I caught his daughter's attention and inquired, "Are you here to take him home?" Before she had made her reply, I had the earpiece planted in my ears and was pulling apart the back of the old man's hospital gown.

"Yes, Doctor," she hesitated, waiting for me to finish. "We're also here to ask a few questions."

Once I had finished listening to his lungs and heart, I announced, "Fire away," and proceeded to palpate his abdomen and check his reflexes. I was mistakenly lulled into a good feeling, thinking this discharge would be easy. I could then get to the office with moments to spare. Instead, I was angrily besieged by a barrage of questions from his sons.

It took me off guard. Many of their questions should have been answered previously during their consult with my partner weeks ago. Even though Helen could not defend herself at this moment, I could not imagine her neglecting to tell them all this information during her consult with them. Yet, they were obviously confused about the prognosis of their father and what his future treatment would entail. It was an awkward situation made worse by their hostility. A furtive glance at my watch gave me a sinking feeling. The patients at the office were going to be seriously annoyed by my delay, but I was trapped.

Embarrassed by the outbursts of his children, Mr. Worth himself sputtered rebukes and angry retorts at his children, raising his bony fists threateningly and adding to the general confusion and commotion. This stirred up more trouble with his sons who shouted down their old man while they continued to vent their frustrations at me.

Within a short time I recognized the problem. The sons especially were spurting question after question at me without waiting for my responses. As a result, they had no concept of what was going on and seemed unable to understand even my simplest answers. I might as well have been talking to a wall. After precious minutes of trying to answer each question as asked, I decided not to waste any more time.

"Hold on a minute, now." I was aggravated and let it show. "Keep your voices down—-we're in a hospital here! Okay, now. I think it would be best if I explained everything to each of you one at a time, you first," I pointed deliberately to the daughter. She had not been so openly hostile during the interrogation. I gambled she would help calm things down by beginning. Turning to the men, I said, "Let her ask me the questions first. Otherwise, it's too confusing for all of you to keep shouting questions at the same time and expect to get a decent answer from me. You can sit here and listen to the answers I give and think of the questions you want to ask while she is speaking. Pay attention to the answers I give, and I promise you by the time I finish, many of your questions will be answered."

This order seemed to work. The men leaned back in their chairs, crossed their arms, and remained silent.

Pointing to my watch, I added, "I can only give you ten to fifteen more minutes now because I'm expected back at the office for my patient hours. I don't like keeping them waiting. First, I will briefly outline your father's overall prognosis and general plan of treatment, then you can ask me your most important questions."

Despite the time constraints, I carefully explained the particulars of his case. Finally, the daughter began with the most important question. I answered as simply as possible without leaving out crucial details. My stomach was tied in knots, but I affected a congenial facade while the interrogation continued. After a good forty minutes—much longer than I had expected it to be—each of them had had a turn, and the questions ended. Apparently, they were sufficiently satisfied with the information I had provided.

Another glance at my watch told me the bad news: 10:28 A.M. I was now over a half hour late for my first office appointment, and I could hear my name being called by the hospital page. "I've got to end this now," I stood up abruptly and headed for the door. "I have to answer that call and I'm very late for my office patients. If you have further questions, you can ask Doctor Delta at your next appointment with her." My adrenaline drove me speedily from their room and back to the safety of the nurses' station. As I expected, upon returning the call, I found that my office was wondering where I was. Jotting my notes quickly in Mr. Worth's chart, I wrote an order discharging him, made a mental note to tell Helen what had happened, and then shot out of the hospital.

I was trembling from the aggravation when I started the car. It was a lousy feeling and made me question whether it would be noticeable to the patients at the office. Although the drive to the office was not long, I hoped it would be enough time to make peace with my nerves before I saw my first patient.

Things had to change. With the growing number of hospital patients, the occasional altercations with irate family members, and the tension to get back to the office by a certain time, it was becoming nearly impossible to make rounds at the hospitals. In addition, I was concerned that my patients were as much victims of this game of "beat the clock" as were my nerves. Since there were five of us in the practice, it would make sense to separate days-on-rounds from days-at-the-office and perhaps relieve the problem. While driving hurriedly to the office, I made a mental note to suggest this at the next office meeting with my partners. It was long overdue.

* * * *

By the time I arrived at the office, forty-five minutes late for my first appointment, I had transferred my anger to nervous energy. I charged in the side door, whipped off my jacket, scarf, and gloves, then donned my white coat. My partner Michael Kruse, who shared the office this day with me, nodded a warm but fractured hello as he quickly disappeared into an examining room. He too was busy with a heavy patient load. Meantime, I shouted the names of several doctors my secretary would have to call for me during the day, reviewed the list of twenty-two patients, and checked the waiting room for familiar faces. About three or four patients were ready for me in the examining rooms. With the help of my nurses who had set things up in advance, I was able to go immediately to the first examination.

Between the phone calls and the patient load, the office was hectic. Somehow, I managed to recoup my time with well-patient examinations. Apologizing for my tardiness and for the brevity of my office visit with each of them, I explained with all honesty that time restrictions forced me to hurry. Most understood.

By 6:30 P.M., I was seeing the proverbial light at the end of the tunnel. It was with a sigh of relief that I entered the room in which Carl Drake sat.

"Carl!" I smiled, "It's good to see you." I remarked honestly as I seated myself on the stool next to the medical cabinet. Then momentarily, I felt apprehensive. I was assuming he had returned for a follow-up visit; what if he had a relapse of his fatal melanoma? After such a whirlwind of a day, I couldn't cope with another disaster at that moment. "How are you feeling?" I asked cautiously, bowing my head to check the latest report on his blood work. I almost couldn't look at him. My mind raced in fear of his answer. It was only thirteen months since our phone conversation late that November night. What if misfortune had struck him so soon? Urgently searching through his blood work for bad news, I held my breath.

"Exquisite!" he answered happily in his own unique way. "I think Lady Luck is still on my side."

I let out a sigh and chuckled at his amusing reply. His positive attitude certainly made him a pleasure to see. The results from the blood work in his chart partially confirmed his statement, but he'd need a thorough physical and an x-ray before I would truly feel comfortable. I leaned back against the wall in relief and asked. "So, what have you been up to lately?"

He gladly chatted about the good things in his life, his girlfriend, his latest sport activity, his job with a nuclear power plant under construction, while I silently worried about finding any suspicious blemishes, lumps or enlarged organs when I examined him.

"Well, let's examine you and get you out of here so you can get on with your life." I offered after I had rested a bit.

Although somewhere I had found the extra energy to listen to him, I realized I was physically and emotionally drained. Fortunately, he was the last patient of the day, and as I had hoped, his physical examination was without any unwanted discoveries.

"Anything wrong, Doc?" he questioned me as I silently returned to the counter to write on his progress sheet, "Is it me or are you having a rough day?"

"Oh no, no. So far things are checking out fine, but I am going to order an x-ray. No, it's certainly not you. It has just been a long, hard day, and I am beat!" I didn't have the strength to say anything more. After I had checked him thoroughly, I sent him out to get x-rayed. Minutes later, the technician placed the fresh film on the view box at the nurses' station. My examination of the film once again proved everything was within normal range. I was extremely relieved.

"I won't have to see you for another few months," I informed him moments later, "unless, of course, you have a problem and wish to have me check it out. Otherwise, let's say you can come back for a follow-up visit in three months. Do you think you can handle that?"

"Sure, Doc. No sweat. Truthfully, the longer I can stay away from you the happier I am, if you get my drift. Nothing personal." He winked at me as he rose to leave.

"I understand completely," I smiled back. "So? What are you waiting for? Get outta here!"

"Oh, yeah, one more thing." He hesitated at the doorway, and suddenly my heart skipped a beat. What was this last "little" problem I was going to be hit with now? "It's a little early, but Happy Holidays!" he said and trotted down the hall.

"Oh, that's right... thanks!" I replied feeling almost sheepish for my mild annoyance. "Same to you," I called after him.

The day was virtually over, except for the usual paperwork and reviewing of x-rays. The faces of the nurses and my partner all looked fatigued. Carl and another patient were already bundled up and stepping out into the frigid night air. Calling out their good-byes, some of the nurses were stuffing their arms through the sleeves of their own jackets as they readied to leave.

As tempting as it was to follow their example, I returned to my office and began the tedious part of the evening. The chair looked inviting as I sank wearily into it and started sifting through the piles of paperwork. The tranquility that settled over the office once everyone had gone promised me some time to handle the business concerns of the practice without interruptions from patients, secretaries, nurses, or phone calls. Quickly I sorted out the important letters and reports from the junk mail and brochures.

After a few hours, mental fatigue got the better of me. It was time to quit. While deciding whether I would be able to come in early enough the next day to finish the less urgent matters, I caught sight of a bulletin from Medicare I hadn't seen before: more fee freezes across the board, including lab reimbursement cuts up to forty percent; and there would be no additional billing allowed to Medicare patients for lab fees. We either accepted Medicare's fee schedule totally or we got nothing. We would no longer be allowed a choice in this matter. The sense of outrage I felt was too much. I had had enough strife in one day.

Cursing loudly, I crumpled the letter into a ball, and dropped it on the desk as if it were too hot to touch. I didn't want to deal with any more tough issues.

I decided to go home. As I grabbed my coat, I felt very sad. I felt my whole profession was being undermined by bureaucrats who knew nothing about what it really meant to be a doctor, or what aggravations we faced daily while dealing with disease, or what sorrow we felt when watching someone die, or what hostility we experienced when family members felt helpless about their loved ones and directed their anger toward us. There were nuances to hands-on care of the sick about which no lawyer could make rules or regulations, and there were complications in diagnoses that defied the D.R.G. (Diagnosis Related Group) system.

I closed the office door behind me and breathed in the cold night air. It was tempting to think about not coming back, to leave medicine and change my profession, to do anything else but deal with sick people. I didn't want to think about anything sad anymore. I wanted to feel happy.

And in the quiet of the December night chill, I had a vivid recollection of something amusing which happened to me an evening on rounds not too long ago. I was visiting a young woman in her early thirties whom we had been treating successfully for breast cancer. Despite our success in controlling her cancer, however, high fevers and a nadir low white count had brought her into the hospital suddenly where she was immediately started on IV antibiotics and fluids. Although this reaction to adjuvant chemotherapy was rare, it was not critical, and it would only take about a week before she would be back on her feet and feeling physically fine.

"Good evening, Patty," I had greeted her good-naturedly when I had walked in her room. Dinner had already been served to her, and now finished, it was stacked neatly on the tray. An assortment of magazines were piled near a floral arrangement on her nightstand, and after a long day of bed rest, her sheets and blankets were heaped like low hills across the expanse of her bed.

Immediately, I could see she had been crying, but before she returned my greeting, she wiped tears from her eyes, quickly blew her nose, and attempted to hide her face. "Oh, hi, Doctor Berger," she mumbled sadly.

It was obvious she was depressed about her condition and I couldn't blame her. She was an attractive young woman who, until she developed breast cancer, was pursuing a rewarding career in financial advising. Her world had been turned upside down in just a few months, and now she had landed unexpectedly in the hospital, feeling wasted and extremely scared.

I felt bad for her. She needed someone to take some time with her to console her and remind her she would soon be okay. "I know it's rough..." I began, and without giving it much thought, seated myself at the foot of her bed atop a pile of blankets.

"Oh! Doctor Berger!" she shrieked with her eyes wide in horror.

"What?" I replied startled.

"You're sitting in my bedpan!" She blurted out.

"What!" As I heard her words, I felt my coattails absorb, like litmus, the contents of her full bed pan. It had been concealed discreetly under some sheets.

Jumping up in surprise, I swiftly removed my saturated jacket, stared at it for a split second, and then roared with laughter. Her startled expression quickly changed over to giggles and then finally she too laughed as hard. Together, we laughed so long and so hard that nurses came by puzzled and visitors paused to wonder. They smiled in mild amusement while we continued to roar.

Finally Patty managed to squeal out between hearty laughs, "Oh, I'm sorry! I should've warned you!" Tears were streaming down her cheeks as she clutched her ribs.

"Thank God, you're my last patient! " I chuckled as I removed my glasses to wipe my eyes. "Oh, Patty!" I sighed at last, "You don't know how much I needed to laugh like that after a long day. Thank you!"

"Thank you," she replied with a smile, "I needed it too."

Remembering it made me smile again as I drove home that night.

5~

Taking Heart

Early the next morning at the office, I slipped a new stack of medical journals into my briefcase and cringed. I had to take them home to read and my wife wasn't going to like it. I already had a tower of reading material and journals stacking up near my nightstand. Every time I brought more home, Joan got annoyed.

"Can't we get rid of some of these?" she urged disapprovingly. Our bedroom was becoming overrun with piles of medical periodicals. Joan had used her artistic talents to make our home a warm, enjoyable, and comfortable place to live, and she wanted it to remain in reasonable order.

Whenever she complained, I accused her of being unsympathetic. "You know that I have to read all this!" I muttered back. She understood that it was important for me to keep informed about medical progress, she just didn't want my journals scattered throughout her house.

I appreciated order. In fact I was an obsessive compulsive about it. Especially in oncology, if I did not demand every i be dotted and every t be crossed, there could be disaster. My office nurses and staff knew how difficult I could be unless everything was in perfect order. Most of the time my staff was reliable, so an occasional slipup in office procedure was understandable. However, I would come down hard on them if it happened too often. I guess Joan was no different.

Putting the briefcase aside, I continued to clear away the paperwork that cluttered my desk. The crumpled Medicare announcement lay on the top of the pile as a reminder of my anger the previous night. Coolly, I straightened it out, placed it aside and continued with the other papers.

It would soon be time to start seeing patients and I was eager to begin. Although I could greatly admire the dedication of medical scientists who had made great inroads in the study of cancer and whose research efforts ultimately helped the patients I treated, I felt my talents were here dealing with people. Here in a private practice was where my abilities were providing the optimum service.

I couldn't practice alone. Oncology was too grueling, and the emotional hassles were too great. Alone, a doctor would burn out too fast. I needed my competent partners to help me share the decisions, endure the hardships, and offset the frustrations. It couldn't be done well, in my opinion, any other way.

"Buzz." The intercom signaled. Chris, my secretary, was calling in. I picked up the receiver automatically while I continued sifting through the papers. "Yes?"

"Doctor Mark Frankel would like to speak to you." Chris said.

Mark was a close friend of mine who in recent years had lost both his parents to cancer. Four years ago, when his mother first was diagnosed with metastatic breast cancer, we were still only acquaintances. I was honored when he called on me to treat her. It was rewarding to know that another doctor valued the expertise of me and my partners enough to send his mother to us. However, the trust he put in us increased the tension.

In the medical community where we practiced, it was not only important to serve your patients well, it was also vital to satisfy the other doctors who make referrals to the group. As long as our reputation remained sound, we would be able to continue doing what we were trained to do. Yet, when a fellow physician entrusted his family to us, it was hard not to give just a little bit more, to go the extra yard—as if the patient were a member of our own family because, in one sense, all doctors share the same professional family.

When Mark's mother came to us, however, her cancer was quite advanced. We did the best we could, given the circumstances. Through our intervention, we were able to give her more time with enhanced quality of life. After she had the benefit of all reasonable therapy, we treated her symptomatically and supportively to a quiet, uncomplicated death.

Mark was pleased with our efforts, because as a doctor himself, he did not regard her death as our failure. Rather, he recognized that we had succeeded in giving his mother relative comfort at a time which otherwise could have been excruciatingly painful for her.

Over a year ago, Mark called us again about his father. Our group did the workup and discovered lung cancer. Again Mark relied on us to help his failing father through the rough times. A few weeks ago his father died peacefully.

"Put him on, please, Chris." I said.

"Hi, Roy." Mark greeted. "I just want to say thank you again for all you've done for me and Dad..."

"I'm glad we could help," I answered sincerely, feeling choked up suddenly. During his father's illness, I had kept my emotions in tight rein so that I could perform at my maximum ability and provide the best care. All I felt then was anxious and tense. There wasn't any time to feel emotional, mostly because it was too much for me to deal with my grief and get the job done right. Now, as Mark thanked me, tears came to my eyes.

"Well you made it easy on all of us." Mark continued. After a few more appropriate words, he switched the topic over to a social event we were both planning to attend with our wives, and ended our conversation on an upbeat note.

I was still laughing at the joke he told me when Chris buzzed me again. "There's a family member of Mr. Worth's on the line to speak with you... a Judith Worth, she says she's the daughter-in-law, and has some questions." I glanced at my watch to see how much time I had before my patients were ready for their examinations, I had almost a full half hour. "Put her on," I replied and shoved some of the junk mail from my desk into the trash pail.

When the connection clicked in my ear, I began, "Hello, Mrs. Worth? This is Doctor Berger. I discharged your father-in-law yesterday, can I help you with your questions?"

"Who did you say you were...? Doctor Braggen? Well, I'll speak to anybody who will listen!" she grumbled loudly in my ear. "We all are very angry at the way things are done without anybody telling us—we seem to be left in the dark about a lot of matters, like what are all these tests for? And what are the results? No one has spoken to us yet about what's going to happen..." Her voice was filled with tremendous vexation.

The belligerence of this woman took me by surprise. I felt my face flush with anger at her words. What's more, I knew how totally uninformed she was. Probably a severe lack of communication (which ran rampant in that family) with those members who had interrogated me yesterday in the hospital was the obvious reason, but now I was the butt of her anger and abuse.

At moments like these, it did not matter who was the responsible physician. If any of my colleagues had taken the call, they would have experienced the same hostility directed at them. Unfortunately, she had called somewhat early in the morning, and although I had briefly mentioned the incident to Mike during office hours yesterday, I hadn't been able to tell Helen yet. I was still unaware of how much Helen had probably explained to them before me.

This was not the first time that a relative had confronted me in this matter, and I knew it would not be the last. Usually, if I am not informed of what has happened between the patients and my partners, I proceed cautiously, to find out who is at fault, if anyone. Everybody makes mistakes or forgets to mention something important every now and then, but in our group, my partners and I try to provide safeguards so no one makes a major error.

The woman on the phone was obnoxious, but her reaction wasn't rare. Most people get angry and lash out when cancer invades their lives. Accusation resulting from the misdirected anger of family members is one of the major heartaches in dealing with people and their diseases. These frustrated relatives are not just venting their hostilities towards the doctors, they are actually blaming the doctors for the diseases of their loved ones. Even though the irrational wrath can be understood intellectually, the antagonism to the physicians is real. In turn, doctors appear more and more aloof in order to mask the aggravation they are actually feeling.

Swallowing my own anger as best I could, I interrupted her. "Wait a minute, let me say something here. Yesterday, in the hospital, I spent the good part of an hour with Gerard Worth, his daughter, Bernadette, and two of his sons, apparently neither of them was your husband, dealing with all the particulars of his condition. I don't suppose you have spoken with anyone who was present at the hospital yesterday when your father-in-law was discharged?" I felt my irritable colon rumble with the suppressed annoyance as I continued. "You should speak to them first, and then get back to either Doctor Delta or myself if there are yet any unanswered questions."

"Oh! Nobody told me!" The disgruntled woman responded with some embarrassment, "Everybody has been complaining... I didn't know..." I listened in silence as she continued, "It's an unbelievably trying situation for all of us. We don't understand what all this medical stuff means or what's going on." She was fetching unsuccessfully for excuses until she finally backed off with an apology. "I am very sorry, Doctor. Now, wouldn't you think somebody would tell me of these new developments before I called? It's just like them to forget us, again. I'm going to call Bernadette as soon as I hang up and give her a piece of my mind. Oh, yeah... Who was the other doctor you mentioned?"

"Doctor Helen Delta," I said with uncharacteristic coolness. "You can check with our secretary for her hours. Let me switch you back," I offered quickly and clicked her off. Whereas customarily, I would have shown a little more cordiality to a relative, I just couldn't bring myself to waste further time placating this woman. Since Mr. Worth's treatment was within Helen's jurisdiction, if I were lucky, I wouldn't have to deal with most of that family too often.

Disconcerted, I realized I did not like to see myself respond in this way, even if it was my self-defense mechanism which had kicked in. I came into private practice because I liked people. Acting this way toward someone made me feel uneasy. Certainly some emotional callouses were normal, but I didn't want to be insensitive. I decided I should check with Helen and my other partners to see if they were having the same difficulty with this family. They probably were. That thought reassured me.

"Buzz." The intercom called. "Yes?" I responded.

"We're processing the bloodwork on Jameson now; she'll be ready for you in a few minutes."

"Thanks for the warning. Buzz me when the CBC is done."

There were only moments left now to finish up. Quickly I reviewed the office financial report indicating some delinquent patient accounts. Although most times these past due accounts resulted from either inadequate insurance policies or lax payment schedules on the part of the patients, I was aware that one particular patient on the list probably was a true hardship case.

Jeannette Dolan was an old woman and a widow, who had little retirement funds and virtually no coverage except for Medicare. Since the drug companies require immediate payment for their supplies, she had often found it difficult to pay for the high cost of her chemotherapy.

In a fee-for-service practice such as ours, usually it is the patients' responsibility to pay in full for the drugs and then seek reimbursement from their insurance companies. Otherwise, it would become a costly and constant out-of-pocket expense for our practice if we waited the five or six weeks for every patient's insurance company to reimburse us. However, as in Jeannette's situation, it was not above our office policy to give reasonable credit so that a low-income patient could continue with treatments. Our feelings have been that we can't let the patient die just because he or she doesn't have the ready cash to pay for the drugs.

In Jeannette's case we would probably work out some arrangement where we would be willing to wait for the 80 percent reimbursement. Whatever she felt she could afford to pay to reduce the remaining 20 percent would be up to her. I made a note to myself to speak to our billing department on this matter. In essence, I recognized that we were taking a large financial gamble on the patient's own integrity.

Ever since an unscrupulous patient deliberately avoided drug reimbursement to us, falsely claiming to have full insurance coverage, and then abruptly leaving our practice with a very sizeable unpaid drug bill, I have been afraid that this duplicity could occur again. I felt exploited and betrayed by that patient's lack of integrity and reliability. These were hard feelings to shake. I had never felt like this before and was ashamed of my distrust.

Pushing past such unpleasant memories, I had to remind myself of another patient for whom our trust was well rewarded. Although her hospital coverage was adequate, she requested we give her the time to pay up debts which her health plan would not cover, such as her outpatient drug costs.

Here again, we had to make a judgment call on a person and decided to grant her the time she claimed she needed. Certainly we would not suspend her chemotherapy just because her account with us was in the red. Treating her under these conditions, I could not help but occasionally wonder whether we would recoup the increasing expense of the drugs we had to administer to her.

Even though it took several months before she managed to liquidate some of her valuable assets, she came through. With a sense of pride at her accomplishment, she visited the office to hand deliver a check for almost the full amount. Thinking about her made me feel better. Perhaps I hadn't entirely lost my faith in people or in my own ability to judge character.

Using the last moments of this private office time to tidy up the desk top, I recalled how I had felt last night when I had discovered the Medicare bulletin. I had gotten angry because I knew how important financial recompense was as a motivating factor in the practice.

Practicing medicine day and night is an exhausting vocation. The work day does not end when the office closes, nor is home life a safe haven from the urgent demands of around-the-clock health service. When that phone rings in the middle of the night, jarring one awake with a life-and-death issue, a doctor cannot put off an important decision just because sleep is more desirable. Such an emergency call means that the life or health of a human being is at stake, and it is the doctor's responsibility to be alert, awake, and aware of all the extenuating health conditions of the patient no matter what the hour. Sleepy or muddled thinking can cost a life.

Or, when after a long, hard day, stacks of medical journals await, the physician is reminded that their neglect means his or her own ignorance. So, family matters become secondary while the doctor burns the midnight oil reading. This could have terrible consequences.

Years ago, when I was still an intern, I observed a twenty-two-year-old woman who was being worked up in the hospital for hypertension. She was a very angry and emotionally confused person and I spent a long time with her trying to understand her angst. Even at this young age, her life was already in a shambles, her marriage was on the rocks, and her high blood pressure was severely jeopardizing her health. It became evident that she had deep psychological problems especially with regards to doctors. But the hatred she bore her father was the center of her emotional troubles. He had been a physician who had spent most of his time with his patients to the detriment of his daughter. She grew up distrusting others, lacking confidence in herself, and seemingly angry with the whole world. Most of all she resented her father bitterly. Her loathing left an indelible impression on me.

After my daughters were born, I vowed that my children should never feel that way about me, that I wouldn't let the demands of medicine make them feel neglected. It would have been horrible to discover that my little girls had grown up resenting me as deeply as this young woman did her father.

However, the real demands of an oncology practice have made my promise a difficult one to keep. Compromise has become the word which must best describe the way I have had to deal with the time constraints of being what I consider a good father and husband, and yet feel I was not failing my patients. This tightrope course I set for myself makes me feel like an acrobatic juggler, in an almost impossible balancing act, so I try not to be too hard on myself when I fall short of my goals.

Instead, I have learned to capitalize on moments of "quality time" by having heart-to-heart talks with the kids, playing games, helping them with school projects and homework. For the most part, however, I depend upon blocks of free time such as weekends, holidays, and vacations to reacquaint myself with my children. As a result they do not feel deprived of my company and I feel successful.

Success in oncology is not so obvious. Pleasures from the job are too often dominated by death. While beating cancer is the triumph I hope for with each case, usually I have to accept the more realistic expectations of controlling cancer temporarily. On occasion, I can find comfort in knowing I have helped give the dying patients extra time to spend with their families or to settle their personal matters—time they would not have had if I had not had some success in staving off the disease—before I helped to guide them to a peaceful end. I can even feel content in helping the surviving members feel less haunted by doubts about whether they had done all they could, or whether the patient had been receiving the best medical care available. Even when the patient dies, I help the relatives come to realize that it was a better alternative than lingering in torment.

Especially when the rewards are limited, I enjoy being appreciated for my efforts, recognized for my expertise and respected for my dedication. I need approval of and recognition for my work to motivate me to continue.

Yet, perpetual motivation can be difficult. There is a correlation between how much doctors will extend themselves, at the expense of their families and their own personal lives, and how satisfied they are with their generated income. If pleasures are harder to get from the job, financial rewards can certainly help compensate for the extra burdens doctors have to endure.

Sometimes when the failures seem too great, the rewards seem too few, and the motivation is lacking, then the money issue becomes important. Money is part of the trade-off; it reminds doctors that they are duly compensated to bear these responsibilities and hardships.

This is why the Medicare announcement was so upsetting. Physicians have been watching financial reward slipping through their fingers from health maintenance organizations, legislation, and bureaucratic red tape. The more this happens the more they become angry about being put in a situation which makes it harder to treat patients just as they had been doing or to feel that they have been fairly compensated. As a result, their fuses get shorter, their tolerance grows less, and they find they do not want to give as much of themselves to the practice of healing as they once did.

"Buzzzzz... We're ready." The voice interrupted. "On my way." I replied. Before I bounded from the office, I took one backward glance at my now neat desk. Guaranteed, by this evening, it would be cluttered again.

Rita Jameson was a lovely old woman, as cheerful and jolly as an innocent child. She was stocky with shocking white hair but her skin was smooth and soft. Her happy face always worn a beaming smile and her voice sounded as delightful as a babbling brook. At eighty-two years of age, she was responding well to her anti-leukemia chemotherapy and I had every hope she would be around for longer still.

"Hello, Rita!" I smiled when I entered the room.

"Hi, Doctor Berger. How are you today?" She asked enthusiastically, genuinely meaning it.

"Fine, thank you," I chuckled, amused by her honest interest. "How are you doing?"

"Oh fine, thank you, Doctor Berger!" she replied bubbling with good humor, "I went to a Christmas party at the club because my ankles don't swell anymore, and if it weren't for my arthritis [which she pronounced art-thee-rye-tus] I'd be back doing my volunteer work and all, but I can't do that yet, not until I feel better."

As she spoke, I was reading her blood counts and was amazed. "You know, Rita," I had to interrupt gently. She tended to get caught up in long digressions. "We must be doing something right because your white count is up to 2400 which is the best I've seen it, and your platelets are wonderful. Your hemoglobin is staying at 12 ½ ."

"Ha ha ha, I'm glad you're happy, Doctor Berger." Rita giggled at most everything but especially when I became technical in my explanations. Every time she addressed me she repeated my name. "I tell everybody: I know he's helping me so whatever Doctor Berger says is okay with me. That's what I say, Doctor Berger."

Her complete trust in my craft was gratifying personally, but the way she stated her confidence often amused me. "Oh my!" I laughed as I made some notes on her progress sheet, "It's so nice to have a patient who is this cooperative."

"I praise you so highly, Doctor Berger. All my friends who saw me in Queens never thought I'd make it, Doctor Berger," she replied, "but you did!"

"Did you see this, Ruth?" Happy to share this small triumph, I called my head nurse over. "2400 with all the right cells! She's really doing great."

"That's terrific. Rita's white count is up to 2400!" Ruth announced generally to the other nurses nearby who also murmured happily. We all liked sharing the good moments.

"It's a miracle," Rita added undaunted, talking to Ruth, who listened politely, "My friends were going to Ireland, and when I got sick they were going to cancel," so I said to them, "If anything happens, we know where to get in touch with you," So they went and came back with Holy Water. So I said, "between Doctor Berger and the Holy Water from Ireland, I'll be okay. Ha ha ha ha!" She laughed.

"I'm telling you, I'm very impressed," I admitted as much to encourage her faith as well as to sustain my own hopes. "Your red blood cells are great, too. Has anybody snuck a couple of units of blood in you since I last saw you?"

She giggled, "No, Doctor Berger, it's all your doing."

"Well, I'm very impressed with both of us," I said.

"Oh, Doctor Berger, every morning I ask God to help me and to have Doctor Berger help me." she grinned broadly, displaying short white teeth while her eyes twinkled. "Thank God for you, Doctor Berger!"

"Yeah? Thanks." In the face of such strong belief, I was a bit awed. "I need all the blessings I could get!" I added humorously, exchanging glances with Ruth who grunted in agreement and left the room shutting the door behind her. "Now, Rita, let me examine you."

As expected, Rita's examination went well. "All right, very good," I concluded, "I've got to get some more blood on you, but you're doing great. I'll see you in a month. Take care and Merry Christmas."

"Oh thank you, Doctor Berger. You have a Merry Christmas too, Doctor Berger, and your family too, Doctor Berger. Thank you, thank you." she continued profusely as I exited. She was certainly a boost to my professional ego and a cheerful way to start my morning.

"Who's next?" I asked.

Ruth pointed to room 3. "Milly Mack."

"Doctor Berger," Chris called over to me from her reception station. "Mather Hospital is on line 2 for you. It's about Mr. Vincente."

I felt uneasy. Hoping, against the odds, that the message was not what I expected, I answered the call.

The exact words spoken into my ear were the standard medical terms about the manner and passing of a human life. Responding in kind, I questioned, for the record, the necessary details I would need to close the patient's chart. Even though the hospital would send me an official document, I automatically scribbled the information on a note pad. Afterwards, as usual, I asked about his wife. "Would Mrs. Vincente like to speak with me now?"

"No, she's too upset to talk to anyone right now," came the reply through the phone, "but she knows you are asking for her. She says she'll talk to you later." It was important that someone from our group touch base with the surviving relatives, but of course we wouldn't press them until they were ready.

"Please tell her I will be in touch with her later." I answered. That was all I could do for the moment.

When the brief conversation ended, and I had replaced the receiver on the hook, I held it down a moment longer than necessary before I spoke, "Chris, please remind me later to call Mrs. Vincente." Feeling a sense of relief that the man's suffering was over, I quieted all other thoughts for the moment, slipped the note pad into my pocket, and went to see Mrs. Mack in room 3.

Milly Mack was a moody patient. A woman in her late seventies, she combined the unfortunate vice of being a chronic complainer with the sometimes disarming coyness of a graciously charming old lady. Although she was diagnosed as having recurrent breast cancer with bone involvement, the antifemale hormone used in her current treatment had successfully stabilized her.

When I entered her examination room, she was sitting beside her husband and chatting on and on in her high-pitched voice about someone's dog. George Mack was merely bobbing his bald head in response to her remarks. She seemed animated and in high spirits so I thought maybe she was in one of her better moods.

"Oh, Doctor Berger!" She grinned with exaggerated pleasantry, "It's so wonderful to see you so soon." Her husband threw an unreadable glance toward her and then looked away. He wasn't smiling.

Now I wasn't sure whether she was being sarcastic or amiable and quickly decided to take her remark at face value. "Thank you," I replied as I seated myself on the stool and swung the door shut with my foot. Feeling forewarned by George's silence, I was on guard against her dark mood.

Before I had time to ask about her health, she began a long narration of what she had been feeling since the last examination. Listening as closely as possible, and taking notes, I was glad she provided me with such extensive information, although half of it I had heard during her previous visit. Still, as long as I could glean from the conversation the necessary information to help me better understand the status of her disease, I mustered the patience to listen.

Truthfully, even though she had been a problem at times, I could genuinely sympathize with Milly. When she was feeling well, she was easy to handle, but when she was worried or hurting, she could be very demeaning to me, her poor husband, or anyone else who was present. At her worst, she would pick on the money issue with a vengeance, complaining about how her poverty and illness were making doctors wealthy.

"Well, Milly, from what you have just told me," I interjected after hearing her story one and a half times, "I think I'll order an x-ray after I finish examining you."

"What? But, Doctor Berger," she whined, " You just had me x-rayed last time. And that was just four weeks ago, right, George?" The pitch of her voice rose. Rudely, she nudged her husband's rib cage with her elbow in order to get him to agree, but he remained silent with a frown on his face. Perhaps, he was anticipating her next outburst. "Anyway, it costs too much. You doctors!" She suddenly exploded in an irrational fit, "You all make too much money already. Leave us poor, sick people alone! We can't afford to go through all these unnecessary tests and still pay your outrageous fees!" Her face became twisted and red with rage.

Stung by her attack, I was irked by her accusations especially since I had already discussed this with her countless times before.

Milly's complaint did not come as a complete surprise, however. She merely echoed what the general public had often expressed. Especially when it came to fees, doctors have always been accused of being mercenary, of making too much money. When the government stepped in with cutbacks saying that the allocated percent of the GNP was too great for basic medical care, and funding was denied, the general public didn't understand the repercussions. Instead they still expected to receive all the best benefits and technological advances available at little cost. Without government subsidy, this would be impossible because doctors, along with the general public, were victims of these cutbacks.

"Milly." I replied as calmly as possible, speaking directly to her. George seemed to have shrunk into his seat. "As you asked us in the beginning, we have always taken into consideration your economic difficulties. If you remember, we have accepted Medicare and have arranged a fee schedule with you so we don't strap you at any one time. I don't believe any of us are overcharging you or overtesting you so as to get more than is necessary." I had to give her time to let my words sink in.

Sometimes it was hard to convince patients that what my group charged was not exorbitantly inflated. In fact, we had been conscious of keeping our costs within a reasonable range so that very few would refuse our healthcare because of prohibitive costs. Some people just had to be reminded that we doctors were also participating in the American work ethic; that is, we worked for a living.

Milly was calming down. Since what I had told her was true, she seemed pacified by my reminder. Her wrinkled face softened its expression and the redness in her face paled. "It's just that if I weren't sick, we could be using this money to spend on our retirement, the way we planned: traveling, living well, vacationing. I feel so guilty about this. Getting sick like this ruined everything for George and me." She spoke softly and then looked away. Her lower lip trembled.

Recovered from his embarrassment, George gently stroked her hand. "Sssh, Milly. Don't say that. It's been okay. Just, you get better. That's what I want most."

"George is right. We all want you to get better, and we're doing our best now, so you can have the opportunity to enjoy your retirement." I responded kindly looking at them both. "Unfortunately, you have to look at this as an investment in yourself. If you don't have your health, you can't enjoy your life."

"I know that you were x-rayed the last time you were here, " I admitted. "The reason I want to see another one, is so I can compare the two most recent films to check for any physical reason for all your discomfort." I helped her up onto the examining table to check her. Listening first with the stethoscope and then palpating her underarms, torso, and abdomen, I was able to reassure myself that her condition still seemed stable. At least, through tactile examination, no masses or organ enlargement seemed evident.

"Oh, so you really think it's necessary?" She complained without much conviction. The anger had passed. I realized she was probably now just frightened. Looking toward George for support, she seemed suddenly very vulnerable.

"You've been responding well to the hormone therapy," I reminded them both when I had finished the exam, "but I want to nip in the bud anything else that may pop up." I sat back to write a few more notes on her progress sheet until she was dressed, and then stood up to leave. "Okay, I'm done right now. We'll get you x-rayed and then I'll be back."

At the nurses' station, I ordered Milly's x-ray and went to my next patient. Before I shut the door of room 4, I remembered to make a request to the staff. "Let me know when Doctor Delta calls. I have to speak to her."

Paula Bacus sat quietly in her chair even after I had closed the door. Immediately, I could tell something was wrong. Subtle little things, like the way she sat and the pale color of her cheeks were reliable indicators. Also, she was usually more animated than this. When patients acted this way, it was for a reason.

"What's wrong?" I asked with genuine concern as I glanced at her blood count.

Before she could reply, tears streamed from her eyes and her lower lip trembled. Hastily she dabbed her eyes with a tissue while she attempted to speak. "My leg." was all she could say at first and pointed to a apparent swelling above her left knee.

When I began treating Paula for metastatic breast cancer a few years ago, we felt we had succeeded. She had responded well to therapy and was in complete remission until only eighteen months ago. At that time, the disease surfaced again, this time in her bones, but radiation therapy controlled her pain and a change in her systemic treatment controlled her systemic disease, promising us further success.

Now seeing her swollen limb left me with a sinking feeling. If I were allowed, I would have groaned my own anguish aloud. Instead, I tried to calm her tremendous anxiety with some hopeful explanations. Upon examining her, it became clear that she either had an obstruction from her radiation therapy or else a deep vein thrombophlebitis to which cancer patients are prone.

"Paula," I said gently taking her hand in mine. "This swelling does not necessarily mean your breast cancer is getting worse. It is not an uncommon complication in patients who have received radiation for cancer to get this clotting of the veins." Patting her shoulder with my other hand as I spoke, I tried to keep her as calm as possible. "What I would like to do is admit you to the hospital where we can treat you best. This would mean you will be on complete bedrest, you will have hot soaks put on your leg regularly, and we will give you heparin to thin your blood and break up the clots."

She nodded silently and gave me a weak smile of appreciation.

"Is your husband here with you today?" I asked.

"No, he couldn't get the day off, but my niece is here. She drove me today. She's out in the waiting room."

"Well, I would like to have you admitted directly from here, so why don't you give Roger a call from our office and let him know where you're going and why. I can have my office call admitting immediately for you to line up a room." Again she nodded appreciatively, her smile growing stronger.

Returning to the nurses' station while Paula made her call, I could not help but think of the other possibilities I had not voiced to her. Another medication recently approved by the FDA (Food and Drug Administration) would actually break up the clots better than the heparin, but I was deciding against it because the side effects she could suffer might only compound her misery. In addition, there was the undeniable possibility that the clot might travel into her lungs causing a massive pulmonary embolus, resulting in death. This danger was all too real. As she was already frightened by her pain and willing to be hospitalized for the proper treatment, I did not need to instill more fear as a tactic to persuade her.

"Doctor Berger, here are the Mack films you requested," the x-ray technician said as she placed them on the counter beneath the view box.

"Fine! Let's have a look." I said and clipped them side-by-side on the view box.

"Who's this?" A familiar voice asked over my shoulder as I was comparing the previous and current films. I turned to see my partner Sam Weis as he too looked at the films. Although we were both sharing the office this day, it was the first opportunity for our paths to cross.

"Milly Mack." I answered and then pointed out some of the lung nodules from the November x-ray and showed him a couple of new ones that had since developed. After having scrutinized them intensely, I felt they were artifactual or incidental, from the technique used, and Sam agreed.

"Doctor Weis," Sue informed him, "Mr. Johnson is ready now in room 6."

"Catch you later, Roy," he said as he strolled down the hall.

After dictating my observations of the Mack film into the tape recorder, I asked Chris to try to ring up the doctor who had treated Dominic Vincente at Memorial, and in the meantime, returned to Milly and George as they waited in room 3. Now, Milly was quiet as George held her hand. Obviously, they were worried.

Rather than panic the couple by reporting details that were too technical for them to understand and which were really nonthreatening, I chose not to explain that a confluence of blood vessels resembled two new nodules. In fact, tracing the blood vessels in that area of the x-ray confirmed that the shadows created on the film were indeed artifactual. This was good news, but should I announce it to them, they would be torn by doubts and worried with questions that had no real significance. The best approach for something like this was not to mention it at all. The only effect such information would have had was a negative one.

"Your x-rays are showing very little change, Milly." I tried to allay their fears with a calm delivery. "You have nothing to worry about. As before, we'll just keep a close eye on you." Personally I was glad of the development for Milly's sake, as well as for my own. I would hate having to face them if things were indeed worse. Now, I merely had to reschedule her next examination, "Come back in a month."

Both of their faces brightened with my words, while Milly squeezed George's hand and said, "Well, George, you can't go chasing after the young ones yet. I'm still here for a while and I know what's best for you, you old swinger."

"Milly, what a thing to say!" George scolded her, glancing toward me with a look of embarrassment. "You see, Doctor Berger, what I have to put up with? Sometimes I think she's too much woman." Then he winked.

I laughed while shaking my head in wonder. I would never be able to figure them out. "Take care," I said with relief as I headed back to the nurses' station. While I checked the patient roster, Chris summoned me to the phone. "Doctor Stewart from Memorial is on line 3."

"Hi, Bob!" I greeted him. Describing as briefly as possible the demise of Dominic Vincente, I first asked a few questions about what initial chemotherapy protocol the patient had been on and then requested certain documentation. While I listened to his replies, I noticed two more of my patients being ushered into rooms, and was signaled that another doctor was waiting to speak to me on line 4. Thanking Bob for his time and assistance, I switched over to the next line to speak to the caller. This physician was a surgeon making a referral. After he gave me a brief history, I thanked him for his consideration and agreed to see the patient upon his recommendation.

When I hung up, I began to feel pressured. It was not yet noon, but I realized it was going to be another hectic day. I felt myself shift into a higher gear just to meet the demands. Returning to Paula, I told her I would see her tomorrow during my rounds, and then hurried back to the nurses' station to have the orders phoned in to the hospital.

As the day progressed, many patients filed through the waiting room, the phones chimed constantly, and the nurses and staff worked diligently to keep the office flowing smoothly. Sam and I saw patient after patient, answered caller after caller, and still, we could not seem to decrease the numbers of waiting patients. Sometimes an unscheduled but necessary procedure on a patient would take longer than anticipated. Other times counseling or consoling a patient and relatives would require extra effort. With each examination, I tried to give as much care and consideration as the patient deserved and as the time allowed regardless of the hectic schedule or my growing fatigue.

By mid-afternoon, I finally managed to discuss with Helen Delta the misunderstanding the Worth family had about their father's treatment. In our conversation, she had assured me of what I already suspected, that they had received a full explanation from her, but apparently had not understood. We agreed we would have to keep the communication open among the partners with regard to this case as well as to document everything we did and said with the patient and his family just in the event future problems arose.

Late in the afternoon, before the billing department left for the day, I discussed Jeannette Dolan's delinquent account with my office manager. When I had finished, I glanced through the glass window separating reception from the waiting room and with regret saw that there were still more patients. I estimated by the numbers present that it would take another two hours before I would be done with the patients. After that, I would still have paperwork to complete. Later, I would have to call home to let Joan know not to expect me at the usual time.

At 7:30 P.M., before he left for the evening, Sam popped his head in my office to say good-bye. Seeing how busy I was at my desk with the piles of medical forms to complete, he gave me a wry smile and warned, "Don't make it too long a night, or you'll have nothing left of yourself when you get home."

"Don't I know it," I replied with a grimace. "I just have a few more things to do. Some of these have to be done before tomorrow. See you later, Sam."

Hearing the exit door close behind him, I realized I was now alone. As I returned to the medical forms, I slipped my hand by chance into my pocket where I rediscovered the note paper about Mr. Vincente. Before she left for the evening, Chris had reminded me to make the call, but when I had called earlier there had been no answer. Rather than delay it further, I dialed. This was a difficult moment for me because my sympathy for the patient was mixed with relief. His suffering had ended, now his family's had begun. Somehow I would have to reassure his widow that he had avoided great suffering by succumbing to an early death. In truth, when a family was deprived of a beloved, especially by sudden death, there was sometimes small consolation in that fact.

After four rings, Mrs. Vincente answered. She sounded tired.

"Mrs. Vincente? This is Doctor Berger. The hospital told me about your husband today." I paused long enough to let her comprehend this much before I continued. "I am very sorry for your loss, but I hope you will feel some consolation in that Dominic is no longer suffering. Now he is at peace..." the words were rolling off my tongue, words which I had used so often before that at this particular moment, I couldn't feel the sincerity I was trying to convey. My inner tension and discomfort were too great. I was numbed by my own personal fears of separation and dying.

After I had finished speaking, I felt hollow. There was nothing left of myself to give. Fortunately, she was not able to read below the surface of my words and thanked me for my sympathy. When the click of her receiver ended our conversation, I could not shake the emptiness that overcame me. Automatically, I returned to the matters on my desk, hoping eventually my feelings would come back.

By 9:00 P.M., I had finished writing Paula Bacus' admission workup to the hospital and was ready to go home. Looking over my desk to see what I had accomplished in the last few hours, I could see the orderly piles of letters and paperwork which now lay neatly stacked and assorted. Just one more letter remained unopened. Exhausted and only subconsciously aware that the stationery was not any kind of business head, I fingered the envelope in hesitation before curiosity got the better of me and I pried it open. Handwritten in neat, rolling script was a personal letter to me. I immediately recognized the name of the writer as the daughter of one of my now deceased patients. The letter began:

> It is hard to describe the great loss and terrible grief we have been experiencing ever since our mother, Jeanne Murray, died of her colon cancer this past October. As you well knew her courage and quiet fortitude throughout her illness, we thought you would appreciate knowing how much she respected your professionalism and valued your compassion during her ordeal.
>
> We are only now picking up the pieces of our lives, but our family decided we were long overdue in sending you our thanks for giving her that extra time with us. These past three and a half years that she was able to share with us, to see her newest grandchild's birth, to be present at her youngest daughter's wedding, we owe in part to your expertise. The rest was in God's hands.
>
> All the members of Jeanne Murray's family would like to take this opportunity to convey their thanks and great appreciation for the time, effort, and compassion you afforded our mother until her death.
>
> Dorothy Hendricks.

It was late, I was tired and the sentiment came right through me. Tears welled up into my eyes while I read the letter not only because of its eloquence, but also because I remembered Jeanne well. She was one of my favorite patients, a warm, lovely woman who dealt with her disease with great strength of character. Still only in her early fifties when she was struck with the colon cancer metastatic to her liver, she was always ready to face the hard truth without needing false reassurances from me. No matter how bad the disease became, she rarely let herself lapse into deep depression or self pity. She made it easy for everyone, including me, to deal with her disease. Even her death was without anger or fear.

I wiped the tears from my eyes when the overwhelming grief abated and rose to leave. It was gratifying to receive this testimony. It gave me a feeling of deep satisfaction that I was doing my job well. Receiving the family's recognition and approval of what I had done for their mother during her illness was a reward I sorely needed. Purged by this open flood of emotion, I also found comfort in the realization that my feelings had returned. Now I was truly ready to go home.

6~

Rounds

Poor Mr. Havelock was becoming sour. His urologists weren't really explaining why his pain was getting worse, even though they made quick, daily visits to his hospital room to check on his progress after prostate surgery. At first, he didn't mind their brief appearances because the surgery seemed to have corrected his problem and his recovery progressed as expected. However, when pain returned, he was anxious but polite and waited patiently for them to give him some explanation. Instead, they seemed too busy and their comments too vague.

Perhaps part of the blame for this unfortunate lack of communication was in the way Mr. Havelock addressed his physicians with his questions. His Danish accent was thick, and his English, although somewhat grammatical, reflected the syntax of his native tongue. Often he would merely ask, "Vhen I cud goot huume?" as if such a simple question would bring more detailed answers about his condition.

Instead, the urologists would reply succinctly, "Maybe tomorrow," reminding him, "this thing doesn't always clear up that fast, but you're healthy enough to go home soon." After which, they swiftly disappeared.

Because Mr. Havelock didn't know to what "thing" they were referring, his frustration increased.

The blame just as easily lay with his physicians who might have been subconsciously biased by his accent, mistakenly believing that he wouldn't be able to understand them anyway. They were too busy to see that he was a quietly desperate man.

I was unaware of this trouble when I parked the small cart with my patient charts at a spot outside his room. It was 10:00 A.M. and morning rounds were underway. "Hi, Mr. Havelock." I greeted the gray-haired man cheerfully as I entered. "How are you today?" Although he had been admitted for his prostate surgery under the surgeon's charge, he was still one of my patients.

A man in his mid sixties, Mr. Havelock had had four rough years with cancer. When it first began, his physicians had diagnosed him with a rectal adenocarcinoma and recommended surgery. Three years later, a second mass was discovered in his pelvis and he was sent off to a major New York City hospital for another operation which failed to document a pelvic recurrence of his cancer. Shortly after, when another CAT scan showed another mass, they again tried to come up with more tissue samples; and once again, their pathologists couldn't find tumor cells in the numerous tissue samples, so they sent him back to his physicians on painkillers and washed their hands of the whole messy business. Without a positive diagnosis, no one would give him definitive antitumor therapy. Despite this unusual shrug-off his case was receiving from the experts, Mr. Havelock remained a pleasant and reasonable man.

Eight months later, with his pelvic tumors slowly growing in size, he was referred to our group. By that time, we discovered that his colon cancer had metastasized to his lungs. This was six months ago.

Most recently, Mr. Havelock complained of tremendous pain in his pelvis and had great difficulty voiding with a decreased stream. It was determined that an enlarged prostate was causing the obstruction. The differential diagnosis was benign prostatic hypertrophy, rectal cancer invading his prostate or primary prostate cancer. Both his urologists and I agreed to have him hospitalized for his acute urinary retention and to schedule him for the T.U.R.P. (Trans Urethral Prostatectomy). This operation would not only relieve the obstructing urinary symptoms, but give us tissue samples to send to pathology. The last thing I knew when I made rounds on the weekend was that the procedure had succeeded, the results of the biopsy were pending, and Mr. Havelock was feeling fine.

Mr. Havelock sat up straight in his hospital bed before he answered my question. "I'd like to line up wit you vhat you people tink you can do for me and vhat you gonna accomplish, cuzz here I literally have no communication." His words had an angry sound, even though his manner remained calm.

This wasn't good. I could see his disposition had changed since the last time I had made rounds on the weekend. Immediately, I assumed that he was growing more upset about his condition, and truthfully, I couldn't blame him.

"Tell me what you're talking about," I replied, puzzled. In the back of my mind, I worried about how much time I had available this morning to give him.

"I've been in here since da dhirteenth, right? Dhat's nine days already." His light blue eyes squinted with annoyance. "Vhich is not bad, I guess, 'cept dhat I came here to stay one day. Dhey open me up a little bit down here and put dis in" he pointed beneath his hospital gown to the catheter tube, "to flush da system and it wuz vorking very vell, but on Sunday, dhere is blood and pain, da pain is vurst dhan ever I got before——and I lived wit da pain of cancer fur years now. Dat pain I cud live wit, dis pain I can't live wit and I don't vant to live wit."

102

"What did your doctors say about the blood in your urine and this pain?" I asked hoping I was understanding him correctly. Even though I was pretty familiar with his thick accent, I didn't want to miss any important parts.

"Vell," he answered with exasperation, "Dat's da trouble. I ask is dhere any ting I shud know? Da doctor says to me, he says, 'no dhere's nutting.' But I vant to know cuzz it's not vorking anymore, and I hear dhem talk maybe do it again." His anxious face and overall confusion were quite readable. "Dhat's about vhere I stand. I don't know vhat dhey're doin'. I knew vhat dhey vere attempting to do, but it didn't vork out. I don't know vhat dhey're doin' today."

"So, let me get this straight now. You say that the operation seemed to have corrected the problem for a few days, and when I saw you on Saturday everything seemed fine, but on Sunday the same pain returned?"

"Dhat's right." He nodded. " 'cept da pain wuz vurst."

"And you're also saying that your doctors haven't explained to you yet why this pain has come back even worse than before and whether they will actually go in again and redo the surgery?" I questioned further. I glanced discreetly at my watch, aware that this visit was going to take time. I tried to dismiss the thoughts of the other patients I still had to see.

"Dhat's da whole trouble in a nutshell!" he said adding softly, "I don't see my doctors much."

"They have notes in here every day," I responded somewhat defensively, although I realized they may not have sat down and talked with him at length.

"Dhey're too fast." he replied gruffly.

Suddenly my name was paged for an outside call. The timing was ironic.

"Excuse me, Mr. Havelock, for just a moment." I interrupted. "I have to answer that call. I'll be right back," and quickly I headed into the hall for the nearest phone.

Mr. Havelock's complaint about fast visits was certainly justified. Though I knew his urologists personally as good practitioners, unfortunately they were guilty of a common fault—something all doctors have done during rounds—running in and out of a patient's room.

I recall when I first witnessed this happen while I was a fellow at Memorial Sloan-Kettering Hospital, I was appalled. I was an idealist then and vowed I would never do such a thing to my patients.

Years of experience, however, have forced me to face this problem more realistically. On the one hand, as their doctor, I want to meet all my patients' medical, psychological, and emotional needs, caring for them, and feeling that I have done a good and thorough job. On the other hand, it is impossible to give extra time or extra effort when there is not enough time in the day or enough physical stamina to accomplish all these things.

It is an effort to listen constantly to the physical complaints of people. Among other things, it tests the doctors' patience and endurance. Every time physicians question their patients with "How are you?" they expect to hear a long litany of problems because listening is part of the differentiating process. By listening to the patients' physical complaints, doctors can make important decisions on how to handle their health problems. However, hour after hour of patient after patient, listening to complaint after complaint, without a break, can put the physicians' circuits on overload.

Physically and emotionally stressed doctors will continue to care for patients' medical needs and provide the maximum medical attention, but they can no longer give every patient extra-special personal attention. The result is this technique of self-preservation: sometimes breezing in and out of patients' rooms and cutting corners by shortchanging the well patients because the sick ones need the attention.

I know I get behind for various reasons. Sometimes, I have to spend extra unplanned time with a patient, or an emergency call takes longer than expected. Other times, too many phone calls with doctors or patients' family members bite into the time I have allocated for rounds. Occasionally, I have other commitments to patients in the office who are waiting for me to return. When racing against these time constraints, I run in and out of patient rooms.

And I feel very guilty. I try not to let it happen too often. By the end of the day, I need to go home because that is, in part, where the recharging process takes place for me. Without the time away from physical complaints by patients, I won't be ready for the next day.

This used to be a challenge even at home. When my daughters would greet me at the door at night, before they would say "Hi, Daddy, how was your day?" they would say "Daddy, my throat is sore!" or "My tummy hurts!" Or my wife would approach me with "I'm having trouble with..." This, after twenty-five or thirty patients have complained to me about their physical ills all day. Even the people I love the most in the world didn't understand that I couldn't hear any more, that I couldn't take it any more, that I had to distance myself for a while from everyone's complaints, including theirs—at least until after dinner.

It was hard enough wearing my medical thinking cap all day. I didn't want to do it at home too. I needed time to relax, to release the anxiety and tensions of the day, have a peaceful dinner, and then go to the necessary tasks of dealing with problems. At least a short rest would help clear my mind. Then I could sort what's unimportant from what's important when my family approached me with their health problems.

Making a new rule solved this problem at home. Instead of bombarding me with physical complaints when I come in the door, my family holds their problems until after dinner. The rule before dinner is to tell me only good, pleasant, entertaining things that have happened to them during the day. Save the problems for later.

With regard to my patients, I can't tell them not to complain. They are supposed to tell me their physical problems. But rather than become inundated, I have sought to find a suitable compromise as well. Extra is the key word here. When I can't give that extra amount of time or emotion, then I try to do just enough, to find the right balance between giving too much or too little. So, if I have to shortchange a patient one day, I will "longchange" that patient the next.

In Mr. Havelock's case, even though I understood the tension which doctors were under during busy rounds, I felt the extra time was necessary. It also bothered me to see rational patients left in the dark about their condition. Here was a prime example of a poor guy—a reasonable man—with terrible pain, who wanted and deserved to know what was going on. It was bad enough that there were often no pat answers, but at least someone should have taken the time with him. Time alone, even without false assurances, might make him feel better.

When I answered the outside page, my secretary Susan told me that Mrs. Carol Birch, a woman in her eighties, would be arriving at the E.R.(Emergency Room). Then Hank got on the line. "She was here at the office," he explained, "for her follow-up, but it's obvious she's in bad shape with a white count of 200. Anyway, her granddaughter will be bringing her to the hospital later this morning."

"Fine, Hank." I replied. "I'll meet her in Emergency when she arrives." Hanging up, I hurried back to Mr. Havelock's room, ready to resume where we left off, knowing I would still have to visit other patients before the E.R. paged me.

The first time I had walked into his room, I had expected to discuss his pain medication and to determine whether I should increase his medication by mouth so that he could get off the injections of morphine. It was supposed to have been a brief visit. However, this second time I entered his room, my priority had changed. He really needed someone to listen to him. I would give him the time now and hope to be able to compensate for it later somewhere.

Seating myself near his bed side, I told him, "Let me look here in your chart to find the surgical note, so maybe we can see what the urologists have said." I suspected from what Mr. Havelock had told me that the urologists were probably waiting for the customary pain and blood in the urine to go away in time as things healed, but since I wasn't a urologist, I didn't want to misrepresent the facts to him. However, for the time being, I was willing to piece together enough information from his chart so I could give him decent answers.

Quickly, I tried to reconstruct the series of events. His doctors had written follow-up notes daily on the progress records. They did mention the blood in his urine, acute pain in his groin, and the prescribed doses of morphine. Finally, I found the surgical report. After scanning through it, I admitted. "Well, I'm not a surgeon, so I may be missing something here. The report doesn't sound unusual. What I will do is talk to your doctors directly. When you have trouble with the plumbing, you call the plumbers, that's why we call the urologists. I'll phone them later." I scribbled a reminder on note paper and tucked it in my back pocket. "From what it says here," I continued, "I'm a little confused about the location of your pain. The pain I had been treating you for and which I'm familiar with comes from around the back. Where is this pain you're talking about now?"

When he showed me the area, I was surprised. He had indicated a different site. "You mean the pain you're talking about is a different pain? It's not the colon cancer pain which was present in the pelvis for a long time?"

"Much different!" Mr. Havelock replied.

"It's a plumbing pain?" I questioned again to be certain.

"Right!" he opened his blue eyes wide as if glad to have such a neat label for his problem. "A plumbing pain!"

"Are you still in a lot of pain now?"

"Da pain medication helps. You see, da original pain started almost better dhan fur years ago after da colostomy."

"That pain I'm familiar with. It started when you had the first pelvic recurrence." I cut him short, I couldn't spare too much time on repetition. By comparison, Mr. Havelock wasn't too long-winded. Other patients of mine have been so repetitive that, especially when I'm in a hurry, sometimes my reaction is to say: Hey! I heard this five times already! I'm not stupid, I heard you! Of course, that response rarely leaches through my professional composure. "That's the pain that came around from the back."

"Right." he nodded.

"When did this other pain start?"

"Originally two munts ago. And it's not a nice pain to live wit. It wuz almost completely unbearable, every minute, every five minutes, every half hour. Dhis is vhy I came into da hospital, cuzz of da plumbing pain. I separate dhis from da original pain of da cancer. Fur a vhile dhere after da surgery it seemed better, but vhen it started again it wuz vurst. I tell you, I'm absolutely miserable."

This information cleared the confusion I was having about the two very different pains and completely changed my first suspicions. "So, it's a different pain from your rectal cancer pain and from what I understand, your surgery was supposed to ease that second pain, but you're telling me it hasn't worked?"

"Dhat's one hundred percent right." He looked relieved.

I leaned back in my chair to think before I spoke again. Rectal or prostate cancer can spread beyond the primary site and invade up into the bladder trigone. The surgeons wouldn't necessarily see that unless they passed a cystoscope into the bladder and looked for disease. Even if they did that, disease could be within the wall, and the surgeons would only see a normal bladder, unaware that the cancer was invading the trigone externally and causing spasm. That would explain why Havelock was feeling pain, while his urologists hadn't reported anything unusual in his chart.

Before I could decide what the actual cause of his pain was, I would need to see the pathology report. The thought that it could possibly be prostate cancer popped into my head again. "Well now, Mr. Havelock, your problem is a little clearer in my mind." For the time being, there was no need to frighten him with my speculation. "There are a few possibilities. One is that this pain you're having could be just bladder spasm as a result of the surgery. The organ got irritable and so the medicine your doctors have you on plus this catheter for irrigation will help get you better. Once that pain eases, we can then concentrate on your original cancer pain." For the patient's sake, I hoped that the pathological report would be "benign prostatic hypertrophy."

Mr. Havelock nodded. "Da odder pain, dhat is sumting I vanted to correct after da plumbing vorks vere completely corrected. Dhat's all. I know dhat I can't go on forever—nutting goes on forever."

"Okay," I continued, silently admiring his resolve. "That's our goal. Two other possibilities exist. If your prostate has been invaded by the rectal cancer, then we can possibly shrink it with x-ray treatments. The other possibility is that you have a new kind of cancer, one that started in the prostate gland. If your surgery didn't get enough of this prostate tumor so that it's irritating your bladder and causing the pain, we can possibly avoid any further surgery by giving you a medicine, a female hormone..."

For the first time that morning, Mr. Havelock chuckled and smiled.

"... that I can give you for a short period of time." I continued wondering what I had said that so amused him.

"I'm gonna buy a pink shirt," he winked at me, his face all smiles.

The image of him in pink was suddenly very vivid, and I chuckled in reply, "You're going to be very attractive, I promise you!"

Mr. Havelock laughed broadly.

After laughing with him, I continued, "This female hormone acts almost immediately to start killing cancer cells and we'll shrink that prostate down to the point that if it's irritating the bladder, it should give you reasonably quick relief." I felt I had accomplished much by restoring his sense of humor.

"It's guud to know vhat can be done. If dhey had told me all dhat I'd probably wud feel better. You've given me time and dhat helps." Even his rigid posture had relaxed; he leaned back in bed and stretched his arms.

"Next time, Mr. Havelock," I advised him, "you should be more specific about what questions you want answered. If you get flustered because they come in so fast, write your questions down ahead of time so that when they do come in, you can maximize the time they have to spend with you. Then tell them how upset you are and that you would like them to please sit down a moment with you to answer a few questions. They're good guys. They'll listen."

"Okay." he seemed content.

"I'll check with your urologists but to sum this up, for now, this is what what I suggest we do: first, continue with the pain killers and at the same time continue with the medicine they gave you. Second, if we don't see relief in a day or two and the urologists say they don't think it is bladder spasm from the surgery, blood, and clots, at least by then, we will have a pathology report and can decide on radiation or hormone treatments, depending upon which tumor, if any, we are dealing with."

"Now dhat sound like you and dhey are going to vork togedder. Before you vere 100 percent separated."

"You're right. It's supposed to be a team effort, but you have to remember, when I saw you Saturday, everything seemed to have been going fine. You're a nice guy, Mr. Havelock. You deserve to have someone sit down like I have done. I'll tell your urologists what we just discussed. This is going to be a collaborative effort, so between you, me, and them, we'll take care of you. Okay, now?"

"Okay, Doc!" he was smiling when I hurried out the door.

Mission accomplished, but my watch read 10:30. I had spent almost a full half hour on one patient. Despite the time constraints, I felt satisfied that I had helped this poor man out of the dark. This was a service I enjoyed performing. Most of all, his earnest and warm appreciation made me feel wonderful.

It was a reward that kept me going.

I rolled my cart down the hall to the next patient's room, selected the appropriate chart from the stack, and read through it outside the door. The record showed that the newest partner in our now six-doctor group, Gregory Tee, had done the original consult. I was pleased. He always did a thorough job without seeming insincere or aloof. Since he had joined our group almost a year ago, I had yet to hear complaints from our patients or other colleagues about his performance. Although bad things have been said about the rest of us from time to time, Greg has maintained good rapport with everyone, especially his patients.

"Hi! Mr. Burr." I greeted the patient, "I'm Doctor Berger, one of Doctor Tee's associates." Anton Burr's file was thin. He was being admitted by his referring physician for an extent-of-disease workup and a treatment plan was to follow once the results were in. Because he was new to our group, this was the first time I had met him.

Anton Burr was worried and pacing the room when I entered, apparently apprehensive about the results of his tests. A good-looking man of fifty-five, he had recently been diagnosed with colon cancer. Standing near him was a younger version of himself, obviously his son.

They both returned my greeting with similar, sad faces.

From what I read before I entered, he had almost completed his diagnostic evaluation. Until I could look at his scans and speak to his admitting physician about my findings, however, there could be no decision on when to begin his treatment plan. Even though we were still in limbo, I wanted to reassure him that we would be coming to a decision soon. "I do understand, Mr. Burr, how hard it is to wait for everything. After all the test results are compiled," I explained, "our group will pull all the information together. At that point we'll sit down with you and give you a full explanation of any therapy which we recommend. Hopefully by tomorrow we should have a decision."

Their faces were tight with suppressed emotions.

"In the meantime," I continued sympathetically, remaining clinically detached at the same time, "we'll be in touch with your docs to see if there is anything else they want to do, okay?" I was feeling bad for them but my protective shield prevented me from delving deeper into the reasons behind their frozen expressions.

Both heads nodded. They understood. Apparently, neither of them wanted to talk about it. Until they were in touch with their emotions and verbalized their fears to me, I had no more to say.

"Take care," I said kindly as I exited. Unfortunately, I couldn't afford to stay and chat to draw them out emotionally. Also, it was still too early along in his case to add to the information we had. My partner's note, written yesterday in the chart, mentioned that the patient had been apprised of his options should results of certain tests be positive, so for today, I had done all I could do without being repetitive. I regretted the brevity of my visit but I had to press on.

Out in the hall, I rolled the cart down to a different room and began reading through the next patient's chart. While I was there, one of the nurses came over and expressed her misgivings about Anton Burr to me.

"I think he's holding a lot in," she whispered with genuine concern.

Listening closely to what she had to say, I appreciated her input. "Well, it's all new to him and very scary stuff," I replied looking up momentarily from the next patient's chart, "and there are still many unanswered questions."

It was always helpful to hear the nurse's point of view. Nurses are the eyes and ears for the busy doctors. What the doctors might miss during their visits on rounds, nurses notice during their full shifts. As they attend to each patient's medical and bodily needs, nurses become attuned to the patient's overall health and can recognize when a patient is suffering from more than just physical distress. Their impressions of patient behavior have often been indispensable to me.

Yet nurses do more than just communicate a patient's anxieties to the doctor. When they have determined that a patient is depressed or unable to cope with the fear of disease, they will take the necessary time to comfort and care for the patient, holding their hands, hugging them, offering the patients tangible gestures of emotional support——essential in the healing process. Especially in a hospital setting, nurses usually give their patients a lot more emotional support than physicians do.

Something I have noticed almost universally about nurses is that they are good at this "handholding." One reason for this is because during regular shifts, nurses have more time to interact freely with patients. These blocks of time are essential in developing a better patient-health professional relationship. The other reason, which I feel is the more important one, is that nurses are not ultimately responsible for the patient's overall health care, as doctors are. Therefore, they can focus on the emotional support and not fear that they will be held responsible for the serious decline of a patient's health.

Nothing is as frightening as being in the physician's position when the life of a patient is at stake. Doctors have tremendous anxiety over this burden of ultimate responsibility and this, in my opinion, is one reason why doctors distance themselves emotionally from their sick patients.

At that moment, Anton Burr's admitting physician walked over to us as the nurse and I were talking.

"Great!" I exclaimed as soon as I spied him. "Your timing is perfect." Immediately I incorporated him into our discussion and asked about his plans for the patient. His input was helpful, providing me with insights on Burr's personality as well.

When I had finished discussing Mr. Burr with the physician and nurse, it was 10:45 and hospital business was in full swing. Now, dietary carts rolled in to collect the late morning trays and laundry carts cluttered the hall. Medical personnel dodged about. Phones were ringing, patients were buzzing for assistance, runners wheeled briskly by, and everything was in constant motion.

Fortunately, my rounds were pretty routine for the rest of that morning. None of my patients was in serious condition and while some were recuperating from minor problems, others were almost ready to go home. Without seriously shortchanging anyone, I tried to keep a steady pace.

A common problem I noticed with some patients was that they were sitting forlornly in their hospital room and moping about their discomfort or discontent. At which point, I would kid around with light jokes or funny remarks and bring smiles to their sad faces. In addition to my medical attention, the proper blend of compassion and humor would usually do the trick.

Throughout the morning, interruptions were frequent. Either nurses and technicians exchanged questions with me about patients, or else I was paged to call my office for one reason or another. Occasionally, I was fortunate enough to meet other physicians who were involved in my patients' cases. It was always important to work closely with all other physicians on a case. Putting our heads together about the facts would often help determine the best course of action.

That morning, I was particularly happy to bump into Mr. Havelock's internist for two reasons: first, because I was able to inform him about three of the options open to us for dealing with the patient's pain and second, because our chance meeting now saved me a phone call later.

It wouldn't be long before the E.R. would be paging me about the arrival of the woman I was expecting. As usual, there was an inner pressure to keep moving and a tension associated with that pressure. The longer I spent on something or someone, the less I was moving. Back in my mind, the feeling was that if I didn't keep moving, I would never finish. If I didn't finish, I would never get home.

Once I had completed patient visits, I headed back to the nurses' station. Rounds here weren't over yet; I still had to write orders and record progress notes in the charts. It was important to summarize each patient's problems, physical examinations, lab data, and plans in their charts.

This is where the records are indispensable in maintaining the continuity of patient care. Even when I haven't met other physicians in the hall, I can keep the communication open to all medical personnel through detailed notes in the patients' charts, so everybody is aware of what's going on.

Continuity and communication are especially important between my partners and me. All our patients depend on it. Yet, it would waste too much time if we were to telephone each other about every hospital visit. Neither I nor my partners would have any time to see our patients if we spent all day informing each other solely in this manner. Instead, accurate records in patients' charts give whichever one of my partners who might see the patient the next day the exact knowledge he or she might need to carry on, to know where we have been, where we are going, and any important conversation that had taken place. The written record helps my partners to continue as if they had been present the day before, and obviously this helps our patients.

The time was almost 11:15 when I sat down. As a reminder a small poster, tacked to the bulletin board above me, read: "No job is finished 'til the paperwork is done."

There was no denying that accurate paperwork made good legal records as well, I thought to myself.

After recording all the necessary information in each patient's chart, checking the Physician's Desk Reference for doses and side effects, or telephoning other doctors to get feedback or to inform them of important developments, I replaced most of the patients' charts on the rack. All but Mr. Havelock's. I had to touch base with either of his urologists before I was finished.

I called their office, and with luck, got the urologist who actually performed the surgery, Dr. Matthew Rue. "Hi, Matt. This is Roy. How are you? I want to speak to you about John Havelock." I tapped my pen impatiently in rapid rhythm against the desk top. "The poor guy is getting a little upset," I continued, "because he's not getting the story as it is unfolding. He doesn't know where he's at. I know from the charts here that Jay's been making rounds for the past few days, so I'll fill you in."

"Go ahead." Matt replied.

Explaining the details I had learned from both Havelock and the chart, I voiced my concerns. "This man may have rectal cancer recurrent in the pelvis, invading the prostate or he may have a new prostatic primary. We've delayed treating him because of his pain and post-op condition and because, as of yet, we don't have pathology's report. He was having preoperative pain for a number of years due to a presacral lesion—-which God knows why they couldn't get a piece of it when they operated on him several years ago. Yet, since his pain hasn't subsided, I'm not sure if this is all spasm or a component of tumor involvement irritating the bladder trigone."

"The T.U.R.P. was successful." Matt replied. "I got enough tissue to get a decent lumen. It's most likely post-op bladder-spasm from the surgery and bleeding, but I can't rule out tumor involvement at the bladder neck. I don't believe there is anything further I can do for him surgically."

"What do you think of giving him a five-day course of stilphostrol?" I questioned, "If it's a hormonally responsive prostate cancer, it should shrink and then he'll be able to void without the terrible pain he currently has. If that doesn't do it, I'd have to assume it's either bladder spasm, hormonally unresponsive prostate cancer, or his rectal cancer. If it's either of these tumors, we would have to keep the catheter in and consider giving him radiation until it shrinks it down. At that point, we would have a better sense of what's going on."

"I'd wait another few days." Matt replied. "By then, we will hopefully have a definitive pathology report. If his pain continues, it's less likely bladder spasm and more likely tumor. A stilphostrol trial would seem reasonable if it's primary cancer of the prostate."

"Good!" I felt relieved to have his endorsement. "One more thing, try to spend a little time with him if you can. He's a nice guy, and I'd hate to see him become uncooperative because of our poor communication." I said diplomatically. "You could tell him next time you see him that we had a good conversation about him and are working together on this. That would make him feel much better."

"I didn't think I was being abrupt with him." Matt answered defensively. "I guess it wouldn't hurt to give him longer explanations if that would help." I understood the slight embarrassment in his voice.

"Very good. It would be appreciated, thanks!" I said and then hung up.

As the receiver touched the hook, I heard my name paged again. Immediately I returned the call. It was the E.R. The patient I had been expecting, Mrs. Birch, had just arrived and would be ready for me soon. I had only a few minutes to wind things up before I would be needed down there.

Before I could close the chart for the day on Mr. Havelock, I had to call pathology. I needed the biopsy result for Havelock before I could write the orders for the stilphostrol.

The pathologist read me the results over the phone. "It's adenocarcinoma." He said dryly.

"I'm not surprised, but which is it? Can you tell me if it's prostate or rectal?" I asked.

"We didn't do those special stains," the pathologist sounded surprised by my question.

"Well, this patient already has a history of rectal-adenocarcinoma," I explained. "and we think it's in his pelvis, but we haven't been able to prove it at the other hospital where the first biopsies were done. With this new tumor in the prostate, we're not sure if it's a primary prostatic cancer or his original rectal cancer invading the prostatic gland. Could you see what you can do about staining with prostatic specific antigen and CEA [a colon tumor marker] to determine for me which of these two tumors it is?"

"Well it helps to have a full history," the pathologist replied, "Sure, we can do those enzymeimmunoassys. We'll let you know when we have the results, " he said and hung up.

It was not unusual for me to take this much time on a problem case until it was resolved. It made me feel good to face a challenge and to follow it through to a satisfying solution. At least for the present, I could clear further thoughts of Mr. Havelock's case from my mind, and get on with other matters.

It was 11:35. Without a moment to spare, I was off to emergency.

* * * *

C'mon elevator! I thought, impatient for the doors to open to the emergency wing. It was only seconds later that the doors slid apart. I bounded out and strode down the corridor toward the double doors that separated the E.R. from the south wing.

Behind those doors was an antiseptic and bright world. Although the commotion in the E.R. appeared calm, everything was happening at an accelerated rate. A screaming toddler thrashed in a nearby crib while an anxious mother assisted the physician. "... spiking fevers... seizures..." were snatches of the conversation I caught as I passed. A portable x-ray unit lumbered behind me toward another patient lying quietly on his stretcher, while a frail, stark-white haired old woman was moaning softly on a nearby gurney. I suspected she was my emergency patient, Mrs. Birch.

At the nurses's station, I greeted the charge nurse who handed me Mrs. Birch's chart. The information in it was pretty skimpy. Since Doctor Holt had access to our office chart on her, I planned to call him for more details. First, though, I needed to examine the patient.

My guess had been correct. The charge nurse pointed me in the moaning woman's direction. When I approached, Carol Birch's sunken eyes were staring blankly up at the bright ceiling and she was wringing her hands.

"I don't believe we've met before, Mrs. Birch," I began politely, feeling the usual clinical detachment upon meeting a patient for the first time. I wanted to observe her condition without being absorbed emotionally by her situation. "I'm Doctor Berger."

I clasped her frail hand which she had raised during my introduction and held it for a moment while she replied weakly, "It's very nice to meet you." I could see it was a great effort for her to move as well as speak. Even the way she formed her words indicated she was having trouble with her mouth. Past this woman's sick appearance, I detected a proud, sophisticated, and gentle lady.

"Well, I'm here to help," I said patting her hand gently once again before I put it down to open the chart. There was very little history in it, so I closed it and turned to her for the information I needed, hoping she could remain lucid and cooperative. "What are you being treated for?"

"Cancer." she replied.

"Of what?" I asked patiently.

"It started in the intestines," she sighed.

"How long ago?" I realized I would have to coax the details from her.

"A year," she sighed again wearily.

I could sympathize with the trouble she had responding while she felt as miserable as she apparently did, but I had to keep questioning her in order to help her. The more information I could obtain from her, the better I could determine what care to give her. When I had finished questioning her about allergies she might have or medicines she might be taking, I asked, "How long have you been on chemotherapy?"

"Aaah, not long. I don't remember when it started but about a week or so ago I had my last treatment," she answered softly.

"And what brings you to the E.R. now, fever?"

"Not really. I have no appetite, I'm vomiting everything I eat and I've got a sore mouth," her voice became softer than before, evidently her energy for speaking was swiftly draining.

Taking out my pocket light and clicking it on in the same motion, I said, "Open wide," and examined the stomatitis in her mouth. "Aaah. We'll be able to take care of that problem easily," I assured her, and began the routine for checking her reflexes. By the end of the examination, I had made my decision. "Okay, sweetheart," I comforted her with a soothing voice, "You're going to have to come into the hospital. We're going to fix your mouth, make you feel better, and give you some antibiotics because your white count is very low. We've got to fix you up, all right?"

She nodded.

"We'll make the arrangements to put you up in the north wing; it has a nice view, okay? " I said trying to raise her spirits. "Do you have family with you?"

"Not here," she replied, "My granddaughter had to go home for the children and my husband's at home."

"I'll give him a call for you and tell him what we're doing. Do you have any questions?"

"Can I be admitted right away?" she asked; her voice quivered weakly.

"Just as soon as we finish the paperwork," I assured her. It didn't surprise me that she was willing to be admitted immediately. Her husband couldn't really care for her at home around the clock, and, given her serious condition, the hospital offered her the best chance of recovery.

"Good!" she said with obvious relief, " 'cause I don't want to fool around any more."

"Well, I don't want to fool around with you either," I said, lightheartedly winking at her.

Not only did she laugh aloud at my reply, but the smile remained on her face for a little longer.

I was encouraged. If she could muster the strength to smile, I felt she had enough fight left in her to get her through this. "Take care now." I smiled back, patting her hand once again and returned to the nurses' station. I called her husband both to inform him that we were admitting her and to ask for more details on her condition.

Afterward, I called Hank back at the office. "There is no question, she's dehydrated and I'm treating her for sepsis," I told him, "but I need to find out if she has metastatic colon cancer. Her husband doesn't know."

"Yes, it's metastatic to the liver but according to her chart her liver wasn't enlarged as of last week." After he read further information to me over the phone he added, "She's Mike's patient. We'll let him know what's happening."

"Thanks!" I said and hung up. Meticulously I spelled out my orders in her chart while the nurse stood nearby.

As I gave directions, I greatly appreciated having alert and attentive nurses. There was no way I could have taken adequate care of my patients without their help and willingness to carry out my orders. But what made good nurses stand out was their ability to be compassionate beyond their assigned tasks, providing warmth and affection on an individual basis to each patient. Their genuine devotion was inspiring.

It was ironic that most nurses were grossly underpaid and undervalued for their important contributions to the overall healthcare system. Unfortunately, lack of recognition, appreciation, and financial compensation has caused a serious decline in the numbers of nurses in the last few years. Considering how scarce nurses were, I felt fortunate that this hospital, like the few others in the community where I practiced, was blessed with many good nurses.

Finally, I handed the folder over to the attending nurse and walked out of the E.R. Now it was time for a quick bite. By the time I was driving over to Mather Hospital to begin rounds there, it was almost 1:00 P.M.

* * * *

In routine alone, my afternoon rounds were similar to those of the morning. The practice of reading charts, examining patients, doing the paperwork was the same from hospital to hospital, but that was where the similarities ended.

No patient, no case was ever the same. There were patients whom I hadn't known for very long and with whom I hadn't developed personal or close ties. Some, as relatives of other doctors or hospital staff, were important to me both professionally and by association. Others were patients who were so pleasant and cooperative that I had grown attached to them and genuinely liked them. Still others were nasty, disagreeable, or chronic whiners, and they were extremely aggravating to deal with. Yet, I respected them all, and despite my personal likes and dislikes, acted as affable as possible to everyone as I delivered quality medical care.

"All the world's a stage," Shakespeare had written, and he was right. For most doctors, acting has to be very much a part of the healing process. In some cases, I've had to act positive, for the patients' sakes, about controlling their diseases even though I knew their chances were slim. It wasn't that I wanted to mislead them, it was that I didn't want to devastate them by removing all positive thinking and hope. Gradually, I would let them see that their diseases were progressing, as I had known their cancers would, and bring them gently to the end. This pretense must be precisely balanced. If it weren't, and I was too positive, the disappointed patients and family members would be angry at me for holding out unrealistic hopes and making false promises. If I were too candidly negative about their chances, they would be angry or frightened and perhaps would leave my medical care to seek help from the unproven methods of charlatans.

Sometimes acting would be for other reasons. Occasionally I have had to project that I have an amicable relationship with difficult patients so that they would get better by cooperating with me during treatment. Even though I may have thought they were obnoxious, I still tried to bring their spirits up with humor or answer questions truthfully, but in ways that I thought would make them happy about receiving my care. I would have to put up a protective shield and act as if their antagonisms weren't bothering me. Rarely would I allow my angry side to show, unless I was pushed very hard at a vulnerable moment.

When I was in training, I was very impressed to learn that there was an appropriate time and place to act angry and even yell at patients. Until then, I had believed it was never right to show those true feelings. However, I discovered that sometimes people became passive and uncaring out of fear, or failed to listen to vital information at a time when their lives were literally hanging in the balance over a major health decision—a decision for which success was possible. Then yelling was often an effective tool to bring them around and to prevent them from throwing their lives away.

By the end of a typically long and fatiguing day on rounds, it was very difficult to spend the time, to show each and every patient that I cared. Yet, the patient must always feel that I cared, even when my emotional reserve was low. Acting often helped me continue attending to the medical matters of each patient, appearing to genuinely care, while I preserved the emotions in me for those few patients who may have truly needed special attention.

Although dealing with patients' emotions was often complicated, sometimes it was as easy as telling the patient, "You're going to get better. You're going to get out of the hospital."

It was often the case that patients who were hospitalized and feeling very sick were unsure of what was really happening to them. They tended to imagine the worse, even if things weren't so bad. These patients needed to be reassured. Taking care of their worries and concerns by giving them a perspective of where they were with regard to their disease, telling them that they were going to live through this phase, and that they were going to get better, unburdened the patient emotionally.

However, the range of emotions I experienced, although as individual as my patients, had to be controlled to some degree. This was a must. If I were to give too much to any one patient, even to the ones I personally liked, I would be drained emotionally and there would be little left to give to all the others. Pacing myself against the clock was easy compared with pacing my emotional responses throughout the day to those in desperate need.

This is what it means to be a physician. The practice of medicine has to be above the prejudice of emotion. If it isn't, a doctor may fall short of administering the most appropriate care. His or her own personal biases or fears may get in the way. This is why the act of clinical detachment is such an important art.

Unfortunately, clinical detachment is often not recognized by the general public as a valuable tool in the practice of medicine. Rather, most people refer to it disparagingly: "That doctor is cold, uncaring, insensitive, standoffish..." to quote just a few. From the patients' and their families' points of view, such names seem appropriate.

Years ago, this detachment was termed equanimitas by the medical philosopher, Sir William Osler. It was the ideal of scientific objectivity. By today's standards, however, doctors with equanimitas may be perceived as remote, coolly aloof, or even calloused when informing patients about their illnesses and options. But, without equanimitas, doctors can become too emotionally involved with patients and may be unable to approach them honestly or rationally with options for treatment. Some doctors may even involuntarily overlook signs of progressive disease or may not want to admit to themselves that the treatment failed; then the feelings of loss and failure would be especially painful. (This is why physicians should not practice medicine on members of their own family.)

That is not to say "don't get attached," but doctors must know what the balance is for themselves, so that they will function not only to the best of their abilities, but for the patient's benefit as well. This balance is not the same for every patient nor for every doctor. It changes. Each doctor must use an internal scale to find this important balance.

Some patients need more emotionally from physicians than others. It is very individual. Yet, without the proper amount of emotional support—whether great or small by comparison—from their doctors, patients will not be happy. They will become untrusting and uncooperative. This colors the entire relationship because the physicians feel disliked and then may become defensive.

Doctors have to meet each patient's needs within the framework of their own self-protective (and also the patient's protective) mechanisms. The extent that this crucial balance—which is dynamic and changes over the life of the relationship—succeeds also determines the success of those doctor-patient relationships.

It is not easy striking an even balance among warmth, caring, and clinical detachment, or mixing honesty, objectivity, and hope, but as a doctor, it is what I strive to do each and every day.

7~

Parallels

The Friday Night Meeting was about to begin. First Greg, then Sam filed into the room. Linda, our office manager, sat by the phones at reception while a fresh pot of coffee dripped and hissed at the snack counter, filling the suite with its aroma. Nearby, in half-empty boxes, a few cookies and crackers were left over from the day. I had just returned from rounds and hurried over to reception, where I picked up the Friday-Night-Meeting folder, and then followed Mike down the hall to the room. As we entered, Sam was telling Greg a joke. Moments later Helen and Hank walked in talking about a patient as they sat in their customary seats. It was 5:58, and everyone was assembled.

"Greg's on this weekend?" I asked as I flipped through topics on the agenda in the folder. He answered, "yes." A couple of partners teased him good-naturedly about the routine headaches of a weekend on call.

Communication among associates had been the main purpose of these meetings since the partnership began. Initially, we started these meetings to review with each other the courses of treatment and future plans for all the patients in the four hospitals of our community; there was a tremendous amount of information to cover or exchange. It was especially important for the doctor on call that weekend to be informed because he or she would be seeing the thirty-five to fifty patients a day in the hospitals as well as taking all the phone calls, emergencies, and consultations for the whole group. The responsibility could be very trying for the one doctor.

Over the years however, these meetings did not only bring us up-to-date on the status of all our patients, it also became a forum for disseminating new medical information, for filling each other in on gossip in the hospitals, for vocalizing and settling differences between group members, for deciding on hiring and firing of employees, and for many other practice-related matters.

"Who are the patients Greg's going to see this weekend at Mather?" I asked in general, anxious to get the ball rolling. The partner who normally rounded at

119

Mather on Fridays began filling him in on the particulars of each case, while I sorted through the material in the meeting folder.

Our practice, although run as a P.C. (professional corporation), is in actuality a marriage of six people with different needs and personalities. Similarly, we've had moments of disputes and discord balanced by times of bliss and harmony. Because our professional and financial futures are so closely intertwined with one another, we've learned through these years together—after painful trial and error—to keep our disagreements on as high a professional plane as possible and not resort to petty or degrading comments or insulting innuendo. Although each of us at one time or other has prudently conceded to the occasional whim of another partner, compromise has been a necessity in maintaining an esprit de corps. This allows the group as a whole to perform at an optimum level.

Most of the time, we work very well together. When we're at our best, we're a solid body of knowledge and experience, preventing each other from making errors in medical judgments and helping our patients to the best of our combined abilities. We have six medical minds to file through vast information and to ensure we are abreast of the latest medical technology and developments. When we're at our worst, however, we feel the strains of six different wills and ideas clashing on matters of business operations and personal opinions.

Coming together at these Friday Night Meetings has become a weekly test of the acquired skills at mediation and compromise for us all. In fact, for me, the meeting is one of the most stressful times of the week. After we've all worked a long hard day at the end of an entire week, there is a certain pressure to get the meeting's business completed so we can all go home. It is especially urgent for whichever one of us is on weekend call to end the day. Interruptions are constant, either because of the number of cross-conversations among the partners about the numerous events that have occurred during the week or from hospital and doctor calls that frequently ring in during the course of the evening meeting. Everyone must wait while the one doctor on call handles the problems. The more the disruptions, the later it gets. After a while, the accumulative delays lower our patience and tolerance, as a result some of our worst disagreements and arguments have erupted here and this is what makes it so stressful.

So far this evening, the meeting was progressing nicely. Sam and Mike were explaining details to Greg about a problem case at Mather while Helen and Hank resumed their previous discussion about their patient. Everybody had something to say.

"Bzzzz," the intercom spat.

"Yes?" I responded.

"It's St. Charles for Doctor Tee." Linda explained quickly. "I'll switch you over."

Greg took the phone to handle the matter. Helen and Hank were still deep in conversation.

"Roy," Sam said while Greg was on the phone, "I have to tell you what happened with Marie Gibson, and why she hasn't been coming back." Mike listened with obvious curiosity.

"Why?" I said slowly, sensing a problem while recalling the patient and her case. I had seen the elderly lady on hospital rounds a few weeks ago. The day before my rounds, Sam had admitted her because she was dehydrated, septic, and febrile. About twenty-four hours later, when I saw her, she had been put on IV antibiotics but was still quite listless. On account of her age and disease, I was fearful that she might not make it, but I never let on to her about it. Instead, I tried to soothe and encourage her, checking to make sure we were doing the best for her while she was in this depleted condition. A week later on rounds, I was gratified by her complete turnaround; the hydration had brought up her blood pressure, improving her kidney function. Her infection was over, and she was on the way to recovery. I told her this in so many words. Within the week she would be discharged.

Sam's remark made me realize that I hadn't seen her since.

"Do you remember what you said to her your second time on rounds?" His question caught Helen and Hank's interest as well, and they both stopped talking. Only Greg continued with his phone conversation.

I was on the spot. I didn't remember exactly what I said at first. I remembered using some kind of common expression to explain my pleasure at seeing her obviously so recovered. As I thought harder, I could picture her light blue eyes, her gray hair neatly combed, even the way she was sitting in bed showed that the vitality had returned to her limbs. She struck me as appearing peppy, perky... What did I say then?

"Oh yes!" I remembered suddenly. I said, "Oh, Mrs. Gibson, you look like you're full of piss and vinegar today." Then, I went on to examine her. She wasn't very talkative.

Only Sam laughed when he heard my answer. I was puzzled like everyone else.

"Well," Sam said as his chuckles died down, "Bill Tyler, her referring physician, saw me in the hall the other day and told me that she was offended by a remark you made to her, something about being full of or made of urine."

Everyone started to laugh. I was surprised and not at all feeling the humor of the moment. I guess my remark could have been misconstrued if it wasn't recognized as just a colloquial comment. Maybe it was not such a common expression after all.

"Not only does she not want to see us anymore," Sam continued, "But she left his practice as well for referring her to such vulgar doctors."

Everybody roared. It was both funny and sad at the same time, and laughing was one way to release the tension.

"Oh, my God!" I grinned sheepishly, "I'm sorry she misunderstood me. I really better watch what I'm saying to people." Although this had never happened to me before in all the years I had been practicing, it made me quite aware that some people were too literal and could take common, idiomatic expressions derogatorily, "You never know how they'll respond." I said shaking my head. "Thanks for telling me, Sam." And I meant it.

I have often said to my partners: if I have done something wrong tell me about it; if somebody is complaining about me, tell me about it; if someone thinks I've said the wrong thing, I've got to know. I can't change my behavior or correct my mistakes unless I know about them. And I'll do the same for you.

I feel it's important to exchange feedback about each of us in the group so that we can change what is wrong. How we portray ourselves to our patients, their families, and the rest of the medical community is linked with our good reputations and how many referrals we receive from other doctors to keep our business operating. Yet, it was equally important to stand united and defend any of us who might occasionally be subject to some outside "bad press."

"What did Bill say?" I asked Sam.

Sam shrugged, "He didn't believe you had meant any harm, although when I tell him what you actually said I figure he'll see the humor in the situation."

"I hope so," I shrugged as well. I couldn't dismiss the blame I felt, but I had to put it into perspective. Certainly, I would try to make sure it never happened again.

Once Greg hung up, the meeting resumed. More patient cases were discussed, and every now and then, someone had a criticism or comment about the way things had been handled that took us all off on tangents. Sometimes the airing of differences became heated. These outbursts were typical of what we each felt at these meetings. After suppressing our emotions everyday of the work week for all our patients, it wasn't always easy keeping tight rein on the anxieties, frustrations, resentments, and tensions we were harboring. So we tended to vent at each other. Sometimes, it could get explosive, but usually, at least one voice of reason would intercede in time: "Hey come on," it would remind us, "let's get on with the business. Everyone wants to go home!" and quickly our tempers cooled.

This night, however, before we could press on with the three other hospitals' cases, we had one last patient case at Mather Hospital to discuss. Unfortunately, I knew another dispute was about to take place because I had a problem that needed to be cleared up.

"What happened last week with Robert Frazier?" I frowned noticeably. "Yesterday on rounds, I found that Frazier hadn't been treated the way I had recommended the week before. I'm upset about this. I thought my notes were clear. He had a cord compression, so as soon as the extent-of-disease workup was completed—which would have been no later than Tuesday—he should have started the protocol I had outlined in the chart." The more I spoke about it the more aggravated I was getting, "Why didn't anybody do it?"

As I expected, there were valid, reasonable explanations for such a failure to comply with my request. One by one, each of my associates had a reply. One said that the patient expressed he was in too much pain to begin any treatment that day. So more pain medication was prescribed and the plans to begin chemotherapy were postponed for one day later.

Another explained that the patient was so skittish and uncooperative that his unwillingness to accept treatment made it seem like we were forcing it on him against his will. Rather than coerce the patient, another reprieve was granted.

A third partner admitted that since I performed the initial consult with this patient and I would be there the next day on rounds again, I should start the treatment.

As far as I was concerned, everyone was passing the buck.

"I realize this patient's uptight," I replied in answer to their excuses. "But I spent the time and energy putting a treatment plan together, explaining to Frazier what had to be done, spelling everything out clearly in the chart for all of you to understand, so that even when I wasn't there, he could have been treated quickly. I expected that you'd all help get the ball rolling faster than you did. I'm disappointed. It should've been a team effort."

I paused. Someone handed me a cigarette and without further thought, I took it. Smoking during these meetings was one of the ways I coped with the stress. I knew it was better not to smoke. In fact, I rarely smoked outside of that room. Having treated patients who have suffered and died from lung cancer would normally be a deterrent, but somehow all the tension and anxieties triggered me. I needed to smoke. It dissipated some of the tension, it couldn't do me irreparable harm since I was hardly smoking a maximum of one pack a month, and it was a vice, I felt, I was entitled to have.

"Not all the information was in..." someone else continued. "Without all the test results, it seemed we were jumping the gun."

"I understand your objections," despite my attempts to see their points of view, I was growing flushed with annoyance. "What's bothering me most, is no one wrote anything in his chart! That's how we correspond with one another so we don't have to say everything on the phone. Or if you didn't want to write it in the chart, call me about it. Tell me one way or other. Leave me a note on the front of the chart or on a sticker to tell me or clip a note to our patient card!"

Another remark by someone about not being able to reach me added fuel to my fire.

"For a whole week?" I flared. "I shouldn't have been left in the dark until I made rounds on Thursday." At this point I couldn't control my temper any longer. "I had to waste time trying to reconstruct what had happened. Here I'm thinking: why did everyone choose not to act on my recommendation? What have I missed? There must be sound medical reason for all my partners to hold off, and then I find it's just a failure to communicate. That really ticked me off!"

I was losing my professional cool. Fortunately, this was in the presence of my professional family. All this time, I could feel my irritable colon tightening, like a rubber band. Most times, it acted up when I was stressed by a situation which didn't allow me to vent my feelings. This was what the practice extracted from me personally. This was the price I had to pay for keeping my emotions under wraps in the face of such tragedy and constant disappointments over my patients. I'm sure each of my partners has paid similarly.

The irony was that even though we confronted each other candidly at these meetings, I really didn't know too much about my partners' emotions with regard to their patients. Unfortunately, there was usually little time or perhaps even interest on our parts to give emotional support to one another after dealing with the overwhelming torrent of anxiety, anger, and tension connected with the caring of our patients. Somehow each of us dealt with this difficulty on a parallel level, alone. I thought it was curious that we seldom talked about it. It was certainly a sensitive topic for me, and so I assumed this was the same for the others, but it still remained one of the unspoken issues among us. We rarely alluded to it even at our very vocal Friday Night Meetings.

Criticisms continued back and forth for a little while longer before we all managed to have our say and return to the real business again.

Over an hour later, despite eight more telephone interruptions, we were almost completely done. Only a few more patient cases at St. Charles were left to be discussed when a very familiar name came up. Carl Drake had come back, and so had his malignant melanoma.

Four days ago, Carl was sitting in the examining room at the office, waiting nervously for his routine follow-up visit. I had been seeing him regularly every four months for almost three years. In that period of time, I had always tried to be there for him, answering any questions he had or allaying any concerns he felt. He had a great attitude, tempered by some justifiable fear, and we had eventually developed a strong and mutual admiration.

Until now, Carl had few symptoms from his malignant melanoma. Although I was prepared to help him when the inevitable happened, he had remained N.E.D. (No Evidence of Disease) for so long that whenever I saw him, I subconsciously blocked the thought that his time was running out. I also wanted to believe the delusion he harbored—that by some miracle his immune system had killed any remaining tumor cells.

"Hi, Carl!" I greeted him with my customary nonchalance as I walked into the examining room. "How are you doing?"

"Not so exquisite," he had answered, slightly out of character. His manly face looked drawn, and he was sitting oddly with his legs and arms crossed while he massaged his forehead.

The fear for his condition, which I held inside for all these years, began to surface. "What do you mean?" I asked hoarsely.

"I... I think it's happened..." Carl stammered as he withdrew his hand from his head. A large mass swelled like a ripening fruit from his neck.

I was speechless as I tried to compose myself. My thoughts raced and scattered, muted by deep frustration and profound sadness. I wanted to yell or cry out—as if I had discovered the tumor on my own neck. Soon he was going to die and I had to take him through to the end.

Although there had been little I could offer him in the way of treatment for controlling his particular cancer, I felt somehow I had failed him, that perhaps I had missed something that would have given him more time.

These doubts shook me only for an instant. Almost immediately I began plodding methodically through all the options and alternatives I had researched over the years and knew without question that there was nothing new to help him until the melanoma recurred.

Now that it had, I could only offer him relatively ineffective chemotherapy or symptomatic and supportive care if it had spread beyond the neck. Chances were slim that his tumor, if inoperable or disseminated, would respond to treatment. Still I wanted to offer him some hope, no matter how small. Neither of us could face the probable, so we clung to the possible. If Carl were game, I was prepared to give it my best shot. First, however, I had to document that the lump really was recurrent melanoma.

"Carl, what happened here? When did you first notice this?"

"I guess you want the truth," he replied with disgust in his voice. "Over a month ago."

"Why did you wait?" I reined in my sudden displeasure. Maybe he had a good reason.

"Denial?" He shrugged, "I thought I may have gotten the bump from playing football with the guys. I thought it was a pimple, a bruise, you name it. I gave it every other explanation than the one I knew it had to be. Anyway," he quipped, "What's the use? I was due to see you in a month, so why come running in like a scared kid when there was little anyone could do about it?"

This wasn't entirely true and he knew it. I was angry now. For three years, I had repeated the message that as soon as he suspected anything, he should call me. The faster we evaluated the problem, the greater his chances of survival and extending his life. It could possibly be controlled by surgery if it hadn't spread. Allowing a recurrence like this to grow for a month was like suicide and Carl knew this.

"Dammit! Carl," I yelled, "You know better than that! All you're doing is making it harder for me to help you. There are different ways we can try to treat the disease when it recurs so you can have more time..."

"I don't want more time!" he snapped. "I can't live like this anymore. I can't wake up each morning wondering what black spot or boil on my body is going to turn against me. I'm tired of wondering and waiting for an internal enemy to strike. To tell you the truth, I'm glad it has finally reared its ugly face," he said.

I was shocked by his vehemence, but I accepted it. His anger swiftly calmed me down. "But, Carl, that's exactly what I'm trying to tell you. We both know it's inevitable. We both know that we are limited by certain factors, but it doesn't have to be hopeless. As soon as you found this, you should have come in, so we could deal with it together. We'd take each day, one at a time, and make the best of it."

He stared at me a moment before he asked pointedly, "Who has this cancer anyway, you or me?"

Stung by his remark, I paused before I answered. "You do," I admitted sadly, "but so do I, by association. It's true you're the one who's going to die, but I'm the one who's going to live with the memory that I couldn't do more to prolong your life. Your suffering is part of my life."

Carl sighed long before he replied in a low voice, "I'd rather have your life dealing-with-death than my death-without-life."

There was no disputing that.

"I face my mortality every time I look at you, Carl." I barely whispered. "I see myself, and my own possible future in you. Believe me, this frightens the hell out of me, too! It's true we're at the most difficult point now, but I won't let you go down without a fighting chance. Work with me."

After much coaxing, he agreed and allowed me to admit him to St. Charles' hospital for a extent-of-disease workup. Results would be back in a few days.

At the meeting, I had to ask, although I felt I knew the answer, just to be certain, if there was anything revolutionary that could be done for Carl. I valued the input from the five medical minds that surrounded me. If they couldn't come up with anything reasonable left to do, then I would have to accept that as a sign that basically everything had been done.

Most cases usually start off with relatively straightforward strategies or protocols. Different cancers require different approaches, but selecting the most appropriate one, while crucial, is usually not difficult. The regimen is tried and the patient frequently does well for a time. After a while, if the plan fails and the patient's cancer progresses, there are still good second-line therapies to try. Further along, however, the choices narrow. Third- or fourth-line approaches that may be suitable for the deteriorating condition of the patient and the resistant tumor are harder to find.

That's when the anxiety is greatest for me. I have to come to a point when I must be prepared to stop. I must admit to myself that there is no longer anything effective to justify the toxicity, and sometimes this transition is difficult to make without help. That's when my partners provide me with the necessary feedback, to assure me that I have exhausted all options and that it is time to stop poisoning the patient.

Individual physicians deal with this differently. Some will never admit defeat. They continue to treat patients aggressively until death. I feel that this frequently causes undue toxicity and ultimately robs the patients of some quality of life in the short time left.

When it is clear there is little hope of halting the progression of the disease, less toxic agents are available, such as hormones or mild chemotherapy regimens. One of these can still provide the patients and their families with a modicum of hope. Even if patients' faith in my ability to heal has dwindled, most patients need to hold on to some hope, no matter how minuscule. I cannot and will not take that from them.

Some test results were in already on Carl, one partner informed me. His malignant melanoma was metastatic to the lungs and brain. The needle biopsy of his neck mass confirmed it was from his melanoma.

I grimaced at this news. Essentially, it was a death sentence. How could I tell him? For a moment, I was relieved I wasn't on duty this weekend. Let Greg have the difficult job of discussing his test results and treatment plan with him, I thought despondently. Seconds later, I felt a tug of remorse. I couldn't allow group protocol to shield me personally from breaking the news to Carl myself. I liked Carl. I had grown too fond of him. It wasn't right to let anyone else to do it. I would drop by the hospital on my way home that night and talk to him.

The consensus of our discussion on Carl was to try whole brain radiotherapy and systemic chemotherapy. None of us was optimistic that this course would effectively control his metastatic disease.

"Symptomatic and supportive care" was their final verdict if he progressed on this regimen. Now that we had discussed poor Carl and decided on his therapeutic options—all in about two or three minutes—we went on to the next patient.

By the end of the meeting, my nerves were shot, I had smoked about three or four cigarettes that I had bummed from the one associate in the group who smoked, my colon was constricting, and I was in a terrible mood. It was time to go home, but first I had to see Carl.

A couple of partners almost flew out of the room when we had finished, calling out their good-byes to the rest of us and Linda as they hurried out of the building. Someone exclaimed, "What a headache I have!" and then asked Linda if there was any aspirin or Tylenol around. Two others started a light conversation as they slowly collected their things before leaving, while I pensively thought how to break the news to Carl.

At the hospital, I knocked on his closed door and slowly opened it, hoping he wouldn't be asleep. The room lights were still on. He was lying quietly in bed, his eyes wide open, staring at the ceiling.

"Star gazing?" I asked.

"Hey Doc!" he replied with a smile, obviously glad to see me. I knew that wouldn't last very long. I felt like the grim reaper. "How are you doing tonight?" he asked.

I hesitated before I replied, "Elegant," My voice was monotone and depressed, betraying my foreboding. I knew I had to be straight with him, and I couldn't keep my sadness from introducing the ominous news in advance of my words.

The combination of my unexpected visit and my dark mood tipped Carl off. "What's up?" he finally asked. He seemed suddenly ashen.

I pulled up a chair, put my hand on his arm, and calmly said, "The CAT scan of your chest showed tumor spread to the lungs."

He sighed.

"Your brain CAT," I continued impassively, "showed tumor spread to the brain."

There was dead silence.

We stared at each other quietly, desperately, teary-eyed, unable to say anything. I was choking up.

"How long?" he mumbled.

"I don't know." I said softly.

"What can we do?" His voice was almost imperceptible.

"Radiation treatments to the brain and chemotherapy intravenously." I responded flatly.

"What are my chances?"

"All or none," I said. Carl was seeking his percent-response rate. It was probably less than twenty percent, four out of five were not responding. I didn't feel like giving him that information. I didn't think it was time to tell him that. "For you, either you'll respond or you won't. Do you really want to hear hard statistics?"

"I guess not." he sighed again and looked away.

"Response or no response, you know we'll be here for you." I reminded him gently.

"I know that," he replied, "and I'm gonna hold ya to it!" he actually lightened up.

"Good, you do that!" I smiled, patted his arm, and slowly stood up. "Doctor Tee will be in tomorrow. He'll explain the chemo and begin treatment then, okay?"

"Yeah, okay." He paused and then added, "Hey, Doc, I really appreciated your coming in here tonight and spending the time to explain things yourself."

"I know you do. I couldn't have had it any other way, even though I was tempted." I admitted. "Get some sleep."

"Good night, Doc."

"G'night." I said over my shoulder. I was unable to turn to look at him again.

When I walked in the door at home that night, after wearily greeting the kids and Joan, I hurried upstairs to the bedroom to change my clothes before my dinner. From upstairs, I could hear Joan scolding one of the girls for a minor transgression. Someone was whining a reply. The phone was ringing, and I felt a strong reluctance to go downstairs again. I didn't want to hear any more problems. I just wanted to be left alone, but I was conscious that the longer I dawdled, the more it would annoy Joan.

Joan had made it clear to me that, at times, she resented my coming home and "disappearing upstairs to the bedroom" for the rest of the evening. Yet, I had my reasons for pulling these disappearing acts. At the end of each day, the children tended to bicker and fight. Worn down after giving to my patients all day, listening carefully to all the problems both emotional and physical, thinking and then acting on the solutions, verbalizing to the patients, their families and other physicians, I was exhausted. I needed quiet time for me alone just to recoup. Even though I had asked the family to refrain from expressing any physical complaints until after dinner time, especially on Fridays, when my dinner was so late, I didn't want to hear any troubles after dinner either.

When I came down for dinner, Joan chatted about some amusing incident, and my daughters were now on good behavior. Ali, my eleven-year-old, patted me affectionately on the head as she spoke innocently about her blossoming interest in modern music. (By her teenage years, Allison would be in love with Heavy Metal music and the tattooed, antiestablishment types that played it. Her entire room would be covered with their pictures. I had never dreamed that the smiling infant I held on my lap while studying for my hematology boards would become a head banger, a Heavy Metal Freak. At fifteen, her goal—a doctor's daughter—was to become a photographer for these rock stars! Intellectually, I knew this stage of separation was important, but amusing as it was, I couldn't wait for it to be over.)

Jess, my nine-year-old, planted a goodnight kiss on my lips before skipping off to her room and to bed. She had always tried her best to please. In contrast, Allison had pretty much done her own thing. This probably stemmed from their sibling rivalry. Jessica always loved to come to my office to look in the microscope (when she got older, she would even help file) and in general paid a lot of attention to me. As daddy's little girl, she always seemed especially proud of me and my work at helping people. This pride sometimes had funny ways of exhibiting itself: One time in a restaurant, when the bill came, she grabbed it before I did. Reading the total aloud, she exclaimed, "Wow! that's really expensive, but you can afford it, 'cause you're a doctor." She smiled, knowing that she had said it loud enough for the nearby restaurant patrons to hear. I was embarrassed by her remark but realized she wanted people to know her daddy was a doctor. This was a child's way of being proud.

Normally, after dinner and some discussion with my family, I would return to my bedroom to read my medical journals. There was always a need to catch up on the quantity of periodicals. I felt obligated to my patients and my profession to keep informed. Anyone who puts his or her life in my hands was giving me a very special responsibility, one which I could not shirk. My patients deserved the most up-to-date, state-of-the-art approach in both diagnostic techniques and available therapy. The only ways to fulfill this goal were to read, listen to audio and video tapes, go to medical conferences, and speak with experts in each area.

I had done this for Carl, as well, but I had little chance of success.

For some reason, that night, I didn't retreat upstairs after dinner, because I didn't want to be alone. I needed to be with my family. I needed to be close without physical contact, to be surrounded by happy voices without listening, to exist without doing anything.

* * * *

Months later, during an informal chat with Carl in my office, he spoke candidly of his life. "These past three years have been pretty productive," he confided. "I scaled mountains I had always wanted to climb, I traveled, I achieved daring feats, I learned sky diving, big game fishing—you've done that too, right?" His face was showing the ravages of his disease, his once thick hair was wiped out by radiation treatment to his brain, and although we had been attempting one treatment after another, his tumor was resistant.

"The fishing," I replied succinctly. I knew about his fishing ventures. Years ago, I had suggested he try it when he was feeling particularly desperate. I had told him it worked for me when I needed to get away. The thrill of going twenty to seventy miles away from the sight of land in the pursuit of large sharks and tuna, watching whales cavorting around the boat was both exciting and relaxing at the same time. It had added to my appreciation for nature and had given me a better understanding of just how small each of us really were, especially as a mere speck upon a vast ocean. "I found that it went a long way in restoring my internal happiness so that I could be more effective in my working hours," I admitted.

"Yes. It made me happy too. Real exciting stuff, all those waves and sharks and gulls..." He was pouring out his soul to me and I was feeling overwhelmed. Still, all I could do for him now was listen and support.

"I approached everything I did as if there were no tomorrows," Carl continued, "because, in essence, there weren't..." He was talking slowly through the constant pain. Each time I saw him I had to increase the dose of narcotics. "Looking back, though," he managed to smile, "I realize I have experienced in my short lifetime what others never do in their whole long lives. I guess, I can say I'm lucky that I was financially well off too, unburdened by family ties or responsibilities. No wife, no kids, so I was able to do everything I ever wanted to do. Sarah... you remember my girlfriend?"

I nodded.

"She couldn't keep up with my fast-paced lifestyle. So we both decided to end the relationship almost two years ago. I kind of lost complete touch in the last year, but I heard through the grapevine that she's getting married. I'm glad for her, too. She deserved a normal life. Now, for me," he concluded, "it's time to pay my dues."

Within the year, Carl died.

More years passed. My daughters were reaching their teens, my private and professional experiences continued to stimulate my personal and intellectual growth. In the realm of the living, we all carried on with our daily lives. Yet, even if Carl had lived longer, this passage of time would not have brought him the quick miracles in the development of cancer research that would have saved him or others like him, although many fantastic breakthroughs were in the works. Maybe it was right for him to embrace his death quickly then. Maybe...

I don't know how I would have reacted were I in his shoes.

As they say, "there, but for the grace of God, go I."

8~

Making A Difference

I couldn't do it without them, I thought appreciatively to myself as I hurried down the hall. I had just ordered blood work, prescriptions, and x-rays from my team of nurses and technicians, and then asked my secretary to page one of my partners at the hospital and also to call my wife to tell her about a medical staff meeting I needed to attend that night. All this as I walked back to my office to conclude a consultation I had started a little while ago.

If my staff weren't capable of carrying through for me, I'd be lost.

"Well, Mr. Arossano," I resumed when I had seated myself behind my desk. "You seem to be a healthy man on exam, but on reviewing your x-rays, bone scan, lab data, and pathology reports, it appears that you do have a problem. The cancer that your urologist biopsied from your prostate gland has traveled through the blood stream and has grown in a few areas in your bones. Fortunately, we do have treatment to arrest the growth of these cells. Atlhough it's not curative, I'm optimistic about your chances of response and hopeful about controlling this disease for some time."

Carmine Arossano was a stocky man in his late seventies, who had been been referred to me for his recently diagnosed prostate cancer. His deep brown eyes under dark heavy eyebrows seemed kind and friendly even though his furrowed brow gave his face a perpetually stern expression. As I spoke to him, he habitually rubbed what was left of his thinning gray hair on a mostly bald head, and nervously looked over at his wife.

Angela, a woman similar to her husband in both age and stocky appearance, was seated beside him. Wedging the older couple between them, two robust sons, built like bulls, created the impression of heavy bronze bookends as their four chairs formed a tight stack in front of my desk. I quickly noticed during the first half of the consult that the two bookends, Robert and John, had a habit of speaking for their parents.

"The standard treatment for prostate cancer," I explained, "once it has spread to the bone is to get rid of the effects of male hormone. Since most of the prostate cancer cells need it in order to grow, this can be done in a few different ways. One way is with female hormone, but I don't want to give that to you because it can increase the risks of heart attacks, phlebitis, and strokes. Since you already have a heart problem, we can eliminate that choice. I just like to mention it so that, in case someone asks you about it or you read about it, you know why it shouldn't be used."

The Arossano family nodded and mumbled their agreement, but no one interrupted my explanation, and I was able to continue.

"Another way is with leuprolide, a medicine which makes the pituitary gland shut off the message to the testicles so that the prostate cancer cells don't get male hormone. This is done with a monthly injection of leuprolide and it's given here when I check you on monthly office visits. Leuprolide stops the male hormone secretion from the testicles. It's comparable to having had an orchiectomy, which is surgically removing the testicles."

"Are there any advantages to surgery?" Robert asked before I could continue.

"The orchiectomy," I replied quickly, "is another way of removing the source of the male hormone or androgens. Both leuprolide and an orchiectomy have been shown to be equally effective, and so the choice of treatment is yours. There is a difference when it comes to insurance, however. Leuprolide which, as I mentioned earlier involves a monthly injection, may be partially or completely covered by medical insurance. The orchiectomy is a one-time surgical procedure which insurance companies usually cover completely. However, if you chose the orchiectomy, you wouldn't need monthly injections."

I paused while Robert and John discussed between themselves the pros and cons of the information I had just delivered.

"There's more." I announced after a short time, and both sons stopped talking immediately. "There's another treatment pioneered by Doctor Fernand Labrie for prostate cancer which involves a drug called flutamide," I continued, "and is used primarily in cases where the prostate tumors have traveled beyond the prostate. What Doctor Labrie has shown is that through combination therapy something called 'total androgen blockade,' can be accomplished, which is better than taking away male hormones from the testicles alone. Various studies across the nation and in Canada are presenting more information which supports this method of treatment. So far, total androgen blockade seems to be better than anything else."

This particular therapy held a special excitement for me because, even though it was still considered experimental, I personally believed in it. My involvement with Doctor LaBrie in recent years and with improving the availability of treatments for advanced prostate cancer in this country happened quite fortuitously, but the events that followed snow-balled into something that has become for me one of my greatest sources of personal satisfaction.

Seven years ago, Seymour Mazer, the father of a friend, was found to have back pain due to metastatic cancer and was referred to me by an orthopedic surgeon. I did a complete evaluation and found that he had disseminated cancer not only to several bones, but also to his lungs. The primary cancer, or source, of these cells emanated from his prostate gland. I didn't have much choice—leuprolide was not available at that time—I had to recommend the standard treatment which was a bilateral orchiectomy, that is, the removal of both testes.

The rational for this was discovered in the early 1940's by Huggins and Hodges, who both won the Nobel Prize for their contribution. They had postulated and later proved that prostate cancer cells were male hormone or androgen dependant. By depriving these cells of testosterone, the major male hormone in the body, the cells stopped proliferating and died. The patients fared quite well because the metastatic prostate cancer could be put in remission for varying lengths of time. Male hormone removal or ablation could be attained by various methods including orchiectomy, the administration of female (DES) hormone, or a drug later to be developed and FDA approved called leuprolide. All these approaches were aimed at getting rid of the testosterone which was produced in the testes, and the one common side effect of them all was almost always impotence.

I spent a good deal of time explaining the diagnosis to Mr. and Mrs. Mazer and their son. I was optimistic that he would do well for months, even possibly years, since the treatment was about 80 percent effective. It, however, was not a cure and I also made this clear.

Concerned about his father's diagnosis, my friend's brother, a dentist, had gone to the medical library to learn more about the treatment of prostate cancer. He found that Doctor Fernand Labrie in Quebec, Canada claimed he had a new treatment which was more effective than the current standard, bilateral orchiectomy. Mr. and Mrs. Mazer flew up to Quebec to give Doctor Labrie's clinic at the Laval University Hospital a visit. That visit was to change my future life as an oncologist.

Mr. Mazer decided to undergo therapy in Canada with Doctor Labrie. I was very interested in why, and so Mr. Mazer volunteered to bring me all the information Labrie could give him.

I studied it carefully. At first, a large portion of this information was geared toward the patients and their families and therefore written in layman's terms. However, I wanted to see the scientific literature supporting Labrie's therapy, and fortunately, Mr. Mazer was able to procure that from Doctor Labrie as well.

This reading was very interesting. Labrie's therapy was based upon his hypothesis called 'total androgen blockade' through combination therapy. As I read further, my excitement increased. His theory purported that only 90 percent of testosterone is made in the testes, and that 10 percent of the testosterone was produced in the adrenal glands. Labrie found that this 10 percent of male hormones was further stimulating prostate cancer cells and allowing cells to proliferate even though the majority of testosterone may have been removed by standard therapy, that is, the orchiectomy.

Now, Labrie was using a drug called flutamide in combination with the orchiectomy or leuprolide to effectively remove all the male hormone from stimulating prostate cancer cell growth. Flutamide blocked the uptake of testosterone at a receptor in the prostate cancer cell. When this occurred, the male hormone from the adrenal gland that reached the cell had no effect on cell growth. If testicular androgen was removed by either bilateral orchiectomy or the drug leuprolide, which inhibited testosterone production from the testes, and the adrenal androgens could not stimulate the prostate cancer cells due to the use of flutamide, total androgen blockade was achieved.

"Let's say this room is a cancer cell," I explained carefully to the Arossanos, "In order for the cell to multiply and divide, male hormone must come into the cell (this room) and sit on this chair, which we'll say for this demonstration is the male hormone receptor. If I'm flutamide sitting in the chair, it's occupied, and the male hormone which has entered the room (cell) can't sit down in the receptor. Since a combination of male hormone and its receptor is needed for all replication, this cannot occur. This is how flutamide works."

"Male hormone can't get in the room at all then?" John questioned confused.

"Not exactly." I replied. "Although male hormone can get into the cell, this room, it can't sit in the chair and give orders. It can't give the cell the message to multiply as long as flutamide sits on the receptor. So we can accomplish total androgen blockade by getting rid of the 90 percent of androgen secreted by the testes through either an orchiectomy or leuprolide, and the 10 percent of androgen secreted by the adrenal gland with flutamide."

John and Robert had some questions, while Carmine and Angela listened trustingly. My enthusiasm for combination therapy made me give many more details about how the drugs worked than I would have given normally. I wanted to educate everybody. However, without a course in endrocrinology, most people not in health-related careers found it too much to absorb all at once. Slowly, I reiterated the parts of my explanation that they didn't understand.

At the time I was first introduced to Doctor Labrie's total androgen blockade theory, I was very interested in learning much more about him and his treatment. I was also amazed that he could claim his combination therapy was effective 95 percent of the time, that it increased complete remissions (disappearance of all visible tumors by physical exams, x-rays, or scans), and most importantly, that it significantly improved survival.

These claims made me leery. Through the years, I have heard of many different treatments for cancer promoted by unethical "doctors" or charlatans that, of course, were only ruses for deceiving patients for profit.

The difference was that Doctor Labrie's therapy, seemed to have a hypothetically sound basis and the drug flutamide was made by Schering-Plough, a reputable pharmaceutical firm. When Seymour Mazer asked if I could help him by getting flutamide, which was not FDA approved at that time, it gave me the opportunity to find out more about combination therapy.

The next day, I phoned Doctor Labrie and although he spoke with a thick French Canadian accent, I was able to glean a true commitment by the way he explained his therapy. There was no hollow sound of the charlatan or fanatic to repel me. My gut feeling was that this man was an honest investigator who truly believed he had made a breakthrough in the treatment of prostate cancer. We discussed his theory, his data, and his approach to Seymour Mazer and patients like him. I became convinced that adding flutamide, which was surprisingly nontoxic, certainly wouldn't hurt and might help our mutual patient. It seemed worth it to add the drug, so I asked Labrie how I could help Mazer and obtain the yet unapproved flutamide in New York.

He replied that he could send me the drug if I became a co-investigator on his United States, FDA I.N.D. (Federal Drug Administration Investigation Number Drug) number. I was willing. He agreed to send me the necessary forms, and I promised to relay all the necessary medical data on Mr. Mazer so that Doctor Labrie could forward it to the U.S. FDA, then we concluded our conversation.

About two months later, all the necessary paperwork had been completed and I received confirmation that I could now legally distribute flutamide to my one patient.

At about that time, two events occurred which prompted my travels to Quebec to see firsthand Doctor Labrie and the institution with which he was affiliated.

The first event was when another patient came to me with metastatic prostatic carcinoma. James Bovine was a man in his mid-fifties who had to be hospitalized for severe hip pain. He was so suddenly incapacitated by the pain that he was barely able to walk. A biopsy was done and prostate cancer was found. Although this disease which is primarily associated with older men, is the most common malignancy in men and is the second most common cause of cancer deaths in men, it can attack younger men such as James Bovine.

After Jim's workup was completed at the hospital, I sat down with him and his wife, Kathy, in his room in order to explain the treatment options. He had been sitting in the bed holding his mixed emotions in tight control while Kathy, seated at his bedside, took his hand in her own and squeezed it tightly. She remained silent, but her eyes were glistening with tears. Although they were devastated by the unexpected news of his cancer, they bravely kept their composure. As I observed them wrestle with their turmoil, I felt an immediate rapport with them.

Yet, in my explanations about treatment, I purposely didn't mention combination therapy. It seemed to me too new and I wasn't sure if I could get flutamide for a patient who had not seen Labrie in Quebec. Jim's severe pain and inability to walk seemed to me to preclude him from starting treatment in Canada as Seymour Mazer had done.

Kathy, a well-educated, intelligent woman, was desperate about Jim's chances and asked all the right questions. As I answered her, I continued to contemplate in my own mind how I could put him on combination therapy. That's when she pressed me with the sincere plea, "What treatment would you give him if he were your own father?"

Without hesitating, I began to explain total androgen blockade, qualifying it with the statement that I couldn't be sure I could get the drug from Canada. Finally, I said I would see what I could do and suggested she call me the following day in my office.

The next morning, I called Laval Hospital and spoke again to Fernand who earnestly wanted to help and promised me a shipment of the drug that day.

When Kathy Bovine called me at noon, she was elated with my news. As soon as the drug arrived, Jim began treatment with combination therapy by orchiectomy and flutamide.

At that time, it seemed like a real gamble to me. Even though Jim would come regularly for treatment and examinations, initially I wasn't sure the therapy was actually working. My optimism was laced with strong doubts because I had yet to experience personally the success rate which Labrie had purported.

Although reading reports on response rates and survival was always important when any protocol was selected, the confidence physicians put into specific therapies depended very frequently on empirical knowledge. Even when the data and actual experiences concurred, there were never absolute guarantees that an individual patient would respond favorably. This was why I was apprehensive about Jim's chances. We were still on the threshold of actual experience for combination therapy, and I didn't want to be having unfounded hopes about his potential for success.

As time passed however, my confidence in combination therapy increased for both Jim and Seymour's cases.

Jim did especially well. His prostatic specific antigen (PSA) levels went down and his bone scans improved. Hematologic problems like anemia improved to normal levels.

This gave me immediate gratification.

Each time Jim and Kathy came to my office, we continued to be very encouraged with his response. Jim began to feel better too, his hip pain decreased, and his ability to walk was greatly improved. Feeling such a dramatic recovery gave both Jim and Kathy a new perspective on life. Their happiness gave me great pleasure, making my efforts seem worthwhile. I was also more curious about Doctor Labrie and his clinic than ever before.

The second inducement to go to Quebec was prompted by Ann Lander's nationally syndicated column. Ms. Landers wrote that combination therapy as advocated by Labrie seemed to be a definite step forward. She mentioned that Doctor Bruce Chabner, Director of the Division of Cancer Treatment at the National Cancer Institute, confirmed this. He believed, she wrote, that there was an advantage to Labrie's approach. Furthermore, she pointed out that U.S. citizens did not need to go to Canada for combination therapy, but could be directed to physicians in the U.S. where the treatment could be used experimentally.

Needless to say, after this article was seen across the country, Labrie was inundated with letters and phone calls from prostate cancer patients and their families. I discovered that I was one of only about fifty investigators in the United States that had affiliated with Doctor Labrie. In the metropolitan area, there were only two of us who could legally obtain flutamide and give combination therapy. I began to see a steady stream of patients who were referred from Laval University via Ferdinand Labrie's clinic. These patients were entered on this protocol, and as data from their treatments accrued, it was sent on to Labrie in Canada.

The increasing numbers made data collation a heavy task. It was impossible to do it myself and I owe much to my head nurse, Ruth Degnan, for her dedication, support, and expertise in compiling the data through the years. I have always appreciated her gifts for dealing with patients and doctors and I especially valued her contribution to this project.

The attention Doctor Labrie received from Ann Lander's article generated great debate among oncologists and urologists treating prostate cancer patients all over this country. I heard reputable people dissent strongly about Doctor Labrie and his theory of total androgen blockade, and I began to have some misgivings myself about my involvement in this project.

The doubts played heavily upon my conscience. I concluded it was now time to go to the source and see for myself if what I had been feeling all along about Labrie, his staff, combination therapy, and Laval Hospital was correct. I owed it to all the patients I was treating as well as the ones who would be seeking therapy from me in the future. I had to make sure I was on the right track. I was not sure what I would do if I were to discover that he was running an unimpressive, unprofessional, sloppy, or otherwise mismanaged operation.

Taking advantage of my daughters' spring break from school, I flew my whole family to Quebec. Spring at Mount St. Anne, the charm of Quebec City, and the excitement of taking the girls to a different country—where Allison could practice her French—seemed an ideal way to spend time with my family and accomplish my primary objective.

I spent an entire day at Laval Hospital. I began by meeting Doctor Labrie in the morning. I learned that he was an endocrinologist rather than a medical oncologist or a urologist. Somehow, I was pleased. It struck me that being an endocrinologist possibly freed him from many of the oncological and urological biases and taboos that inhibited new thinking. We discussed his experiments supporting the contribution of adrenal androgen to prostate cancer cell growth. His methodical reasoning not only made sense to me, but it was stimulating. We then went on a tour where he showed me an impressive laboratory employing well over one hundred scientists, all working on his projects. I later discovered that this was one of the largest endocrinology labs in the Northern Hemisphere. The more I saw, the more I was becoming convinced that Labrie's theories weren't pulled out of midair, but had a strong scientific backbone behind them.

I still had questions before I could completely dedicate my scientific will to his cause. For over two hours, he patiently answered all the criticisms that I had heard before I came to Canada. In my opinion, his reasonable explanations refuted the complaints of the skeptics. Even though I did not always have the data or sometimes even the understanding to evaluate his statements honestly, I decided then that I would continue collaborating with him.

The rest of my day at Laval confirmed these feelings. It became clear to me that Doctor Labrie's entire staff of six or seven other clinical physicians all believed that combination therapy was truly a step forward in the treatment of advanced prostatic carcinoma. The clinic and its facilities certainly had the resources to make accurate scientific observations. Since treatment had few side effects, the only downside seemed the added cost of the flutamide to the patients. Most insurance companies would not reimburse patients for experimental procedures or drugs, and flutamide was no exception.

I left Quebec with a feeling of resolve. I would continue to accrue prostate cancer patients on protocol and send the data to Doctor Labrie.

The data that Doctor Labrie generated with my help and the help of his other U.S. collaborators prompted the U.S. National Cancer Institute (NCI) to fund a randomized controlled study comparing Labrie's combination therapy to one of the standard monotherapies (i.e., ablating testicular androgen only) in over six hundred advanced prostate cancer patients. As I continued to recommend combination therapy, I awaited the results of that clinical trial with extreme interest.

All my efforts were proven worthwhile when the NCI Intergroup study was published in August, 1989, in the New England Journal of Medicine. The results were reassuring. The group that received the combination therapy had a longer progression-free time interval and lived a median of seven months longer than the monotherapy group. This was the first step forward in the treatment of advanced prostatic cancer in over forty-five years!

Furthermore, a small group within the study with minimal metastatic disease had about one and a half year improved survival over a similar group of controls. Results from this observation encouraged earlier treatment, even before the patients developed symptoms from their cancer. This new concept contradicted the old "watch and wait for symptoms" approach which was still being followed by many of the nation's urologists. As a result of these findings, the treatment of advanced prostate cancer patients began to change drastically.

Whereas for many prostate cancer patients the benefits from combination therapy have been significant, James Bovine especially has been a winner. To this day, almost seven years later, he has had no gross evidence of cancer and leads a normal life. I light up whenever Jim comes for his routine follow-up appointments, knowing the years he has been alive and has shared with his family were in part due to my gamble using this experimental approach. His terrific response has given me such powerful, positive reinforcement. Whenever a new prostate cancer patient sits in front of me seeking my help, I can honestly say, "I have truly seen patients who have had great response to this treatment." Giving another patient the hope that he will do as well as Jim—maybe even better in the years ahead—helps me persevere.

"Flutamide is not a cure," I explained to the Arossano family, "but it can help you go into remission. And some patients with well-differentiated tumors have responded for years..."

"Bzzzz!" the intercom snapped, and quickly I answered, "Yes?"

Chris announced, "Doctor Holt is on the phone and your wife says okay about your evening meeting. She won't expect you for dinner."

"See if Hank can hold for a minute, thanks." I had some confidential patient information to discuss. The timing was appropriate, however, because I had already spent about a half hour with this family helping them through their concerns. With my other patients still waiting to be examined, an evening meeting starting at six—for which I expected to be late—and Hank on hold, I had to wind up the consultation.

"Now, is everything that I have explained to you clear so far?" I asked them one more time. Again Robert had a question about the monthly injections of leuprolide. Patiently, I explained it again. "Leuprolide stops the testicles from producing male hormone, flutamide stops the male hormones from the adrenal glands from acting on the cancer cell. By removal of the testicles you are guaranteed that there will be no male hormones produced from the testes."

"So it would be better," Robert concluded, "to remove his testes..."

"Leave my balls alone, will you?" Carmine retorted sharply for the first time that afternoon and immediately his two sons recoiled.

Inwardly, I was glad to see he had some fight. He was going to need it to get through the challenges ahead. "Carmine, the fact that you look and feel good, and we had a normal physical exam, except of course for your prostate, makes me think you should do very well. I'd like you to go home and discuss with your whole family the treatment options. Putting a decision off by one week will not hurt, and it will give everyone a chance to think about the decision. I'll see you next week and will go over any other questions that you have."

Shaking hands as the Arossanos exited, I gave them a cordial good-bye, and swiftly picked up the phone for Hank.

"Sorry, Hank..." I apologized and then got down to business.

* * * *

My experiences with treating many prostate cancer patients spurred my deepening interest in the disease. I began to study it more thoroughly than ever before by reading textbooks and the older clinical trials, and then attending seminars and meetings devoted exclusively to prostatic cancer. As my interest in and knowledge of the disease increased, I began to speak with the drug representatives from the pharmaceutical companies which were actively researching the disorder. Despite the positive results of the study, flutamide was still not FDA approved in the United States.

Schering-Plough, already deeply involved for more than fifteen years and the developer of flutamide, decided to make a major financial commitment in the prostate cancer arena. They hired Burson-Marsteller, one of the largest public relations firms in the country, and formed a Prostate Cancer Education Council, originally consisting of six members.

Aware of my involvement with Labrie and my commitment to this group of patients, Schering-Plough asked me to serve on the council. I was honored to be on a committee with such distinguished people as Doctor David Crawford, Chairman of the Department of Urology at the University of Colorado, who headed the NCI Intergroup Study. The other members were all leaders in their fields of endeavor and all had a special interest in prostate cancer. The main purpose of this learned group was to educate the American public about prostate cancer and try to promote the kind of awareness of prostate cancer among men that woman have had about breast cancer for fifteen years.

Up to this time, most men did not want to think about or be examined for prostate cancer, but with funding from Schering-Plough, the public relations experience of Burson- Marsteller, and the professional knowledge of the members of the Council, we began to make a dent. A press conference in April, 1989, led to substantial national coverage. Then, in October, 1989, Prostate Cancer Awareness Week was announced and supported by the U.S.NCI. Across the nation over fourteen thousand men at over eighty centers were given free screening exams for prostate cancer resulting in more media coverage and more awareness of the problem.

As the data from those screenings were still being tabulated, it appeared that a great deal had been gained from the effort. Annual national screenings during Prostate Cancer Awareness Week would continue and, no doubt, would allow doctors to find more cases of prostate cancer in its earlier and more curable stages. At the same time, the screenings would generate tremendous amounts of important information for future research.

The future was still unfolding for me. I continued to grow more active in the field by subspecializing in genito-urinary malignancies and by becoming an adviser on the board of Patient Advocates for Advanced Cancer Treatments (PAACT).

This group's founder, Lloyd Ney, a former Labrie patient, has worked hard to gather and distribute information about prostate cancer to thousands of sufferers. Together we began to plan future projects such as videos for laymen on the most up-to-date treatments at various stages of prostate cancer. We envisioned forming a Prostate Cancer Oncology Group, which would be a multidisciplinary group of interested physicians who treat prostate cancer patients, developing protocols to answer important questions, heretofore not resolved about the disease. We felt that the necessary funding could be obtained by donations from PAACT members, from pharmaceutical companies active in developing drugs for prostate cancer, and through other grants that may be available. A true sense of purpose was evolving.

The growing need for a support group specifically for prostate cancer patients was part of this evolution. In general, all cancer patients and their families could use support outlets, but prostate cancer patients have their own set of problems related to their disease and the lack of sexual functioning that sometimes results from treatment. The more I dealt with prostate cancer patients and worked through the Cancer Care Foundation, the American Cancer Society, and PAACT, the more obvious this need became. Forming a support group for men with prostate cancer had become an additional goal for me. To date, I am one of a very few oncologists who has taken active steps toward developing prostate cancer support groups in the U.S.

My dedication to this worthy cause continued to open up new experiences for me. Recognized as an expert in prostate cancer, I started lecturing primarily to physicians on treatment of prostate cancer patients and began writing articles for medical journals.

Such involvement in this major contribution to public health has been well beyond my expectations of helping patients on a one-to-one basis. I feel elated to have played even a small role. I am especially thankful that my patients who came to me for this treatment received the advantages it had to offer over standard therapy. Personally, it contributed to my sense of self worth, satisfaction, accomplishment, and purpose.

So when I am asked "How do you do it every day?" my reasons are multifaceted and my answers more complex. The intellectual stimulation and gratification from my involvement in research richly supplements the difficulties and the rewards of treating cancer patients and their families year after year. Participating in the compilation of research data, traveling to conferences and seminars, meeting dedicated professionals who all are striving toward the same goal, gives me a fuller perspective on this war against cancer and makes me feel that my efforts on an individual basis as well as in the larger picture are making a difference.

9~

The Spirit Never Dies

I was about to walk into the new patient's room at St. Charles Hospital early one morning as part of my routine hospital rounds. This new patient was Janice Bond, a forty-two year old woman, who was recovering from a colon resection for a malignant tumor that had unfortunately spread to her liver and abdominal cavity. The surgeon could only take out the primary tumor so that it would not obstruct or bleed later on. No further surgery was necessary, nor would it have been helpful. She was going to die from this cancer. Statistically, her chances of living more than one year were less than 50/50, and five-year survivals for patients with this malady are virtually nonexistent. These were the facts. I knew them well.

My partner, Helen Delta, had already seen her in full consultation and had explained to Janice the nature of her illness and the fact that the disease was beyond the surgeon's ability to cure. Chemotherapy was her only option for prolongation of life. Janice was so devastated by the bad news that little else was said on that first visit.

On rounds this morning I was to answer any questions Janice might have and further discuss with her the nature of her illness and its treatment. Having read her chart, I was well aware of Janice's diagnosis, the extent of her disease, and the treatment plan as defined by my partner. I was not prepared for the unusual reception that I would receive.

"Hi, Mrs. Bond, I'm Doctor Berger, Doctor Delta's partner. She asked me to step in and answer any questions you may have since she saw you a few days ago." I pulled up a chair next to the bed, close enough to extend a hand if it was needed. Janice was an attractive woman who had managed to be well groomed with her makeup meticulously applied despite being in a hospital bed recovering from surgery. She obviously was accustomed to caring about her appearance.

"It's not fair," she moaned, and then choked by emotions began sobbing. My hand went out to her. I barely knew this woman. I had been in her room for perhaps one minute and now her hand was in mine, and I was trying to offer her solace. My mind quickly raced over her plight—anger at being cheated of all the years she had planned on, relationships cut short, fear of pain and suffering, and of the unknown, financial concerns, and the finality of death. I had been through it so many times before; however, each patient's emotional reaction is always different. Only the amplitude, hues, and overtones change from patient-to-patient and day-to-day. My reactions might also vary depending upon my mood, schedule, fears, and anxieties at the time.

I was feeling the familiar frustration with how little I could do at times to change a patient's future. The best chemotherapy response might give Janice a year or two of additional life. She was right. It didn't seem fair—not to Janice, not to her family, and not to me, since I was the one whose craft and expertise she would rely upon for help. Unfortunately, I already knew that current science hadn't given me the tools, the drugs, or the knowledge that day to give her what we both wanted... a cure, or at least much more time. There was, however, one thing I could still offer her: a chance—leave the door open a crack—for hope.

Although I knew the chance was slim that ongoing medical research would come up with an answer in the next several months that could significantly turn around Janice's disease, I offered her this possibility not only to keep her going but also to extricate myself from the tragic situation.

All this time she keep looking up at me through a veil of tears. When I had finished speaking, she stopped sniffling and then both demanded and pleaded, "I want you to be my doctor. Please!"

My mind raced forward. This case was going to be difficult for me to deal with emotionally since there was an immediate bond between us. I really didn't have to take full responsibility for her care. Our group worked on the premise that the doctor who sees the patient in consultation continues the relationship in the office. Mentally, I shuffled through some real excuses for bowing out gracefully. I was too busy with too many other patients. Her case was too sad. She was too desperate. My partner would be upset if I took over her case. I knew I had some way out.

"Okay," I said, "we'll be a team."

The bond was formed. I sighed internally. A commitment 'til death do us part. I could feel the roller-coaster pull away from the loading ramp. I knew the ride ahead was going to be frightening, but there was nothing now that I could do about it.

Janice did not ask many questions during this first hospital visit. She sensed that I cared and would do whatever I could. I guess that is why she wanted me to be her doctor. I did not know what had gone on when my partner had seen her, but I supposed that there was a more formal and professional interaction between her and Helen. Helen had a lot of information to obtain during the consultation, Janice's examination to complete, and a report to render on her findings. All this in addition to discussing the diagnosis, prognosis, and treatment with the patient.

For myself, I often find time constraints make nearly impossible the workload of seeing and attending to all the patient's needs in the two hospitals I visit on rounds. When I added the outside phone calls back and forth from my office staff, physicians, patients, family members, with the frequent paging throughout the day from the nurses, to the procedures such as bone marrow aspiration and biopsies that need to be performed or interpreted, along with the emergencies and inpatient consultations, the amount of time I could spend with each patient had to be balanced. The various forces vying for the physician's time could continue well into the evening and encroach upon his or her personal family responsibilities. While it is possible to juggle all this, one of the indisputable facts of the practice is that time is limited. Giving more time to one patient or to a family means cutting corners somewhere else.

It is difficult enough dealing with the thorny problems of needful patients with life-threatening cancers. When the time factor is added to it, the pressure is tremendous. I am sure Helen Delta spent a reasonable amount of time with Janice and covered all the bases in her consultation. The needs of the patient are very difficult to assess correctly in the very limited interaction we have initially. It is over a period of time and visits that patient and physician get to know one another and learn each other's strengths and weaknesses. Yet a bond of sorts needs to be formed quickly so that a therapeutic alliance can be established.

Patients feel the pressure of time as well. While a certain amount of doctor-shopping can be beneficial, it takes time, a commodity the patients probably shouldn't waste without very careful consideration. Cancer cells multiply and spread as time progresses especially if no effective treatment is being rendered.

Perhaps it was because it was early in the morning when I first saw Janice, or perhaps it was her youth, her tragic situation, whatever the reason, I felt compelled to give her all the time she needed. A relationship existed where none had before, a relationship she would need to help her initially fight but then eventually succumb to her disease. Despite my experience with other patients, neither of us could predict the bumpy course we were about to follow together.

Janice's family was understandably shaken when she later told them the details of the extent of her colon cancer. Like many other families, they felt she deserved the best care available. Several days after she was discharged from the hospital, she telephoned me to explain, "Doctor Berger, I want you to know that I trust you and am very comfortable with the treatment plan we discussed. In fact I'd prefer to continue my treatment with you, but my family wants me at least to go to Memorial Sloan-Kettering Cancer Center for a second opinion..."

"Janice," I cut her off, "I don't have a problem with that. You and your family have to feel comfortable with what we are doing. If a second opinion will be helpful, then I would encourage you to go. I would just like to make sure that you see the appropriate person. Whom do you have in mind?"

"My sister is making arrangements with a Doctor Natale Kenny. She heard that Kenny specializes in my kind of cancer."

"That's correct." I assured her. "She and I trained together at Memorial and she stayed on when I came out here. I will be happy to speak with her after you see her and work with her if she has any different ideas or suggests some modifications of my treatment plan."

"You can do that? Great! It makes me feel so much better that you can work together. I'm sure my family will be satisfied with that too. Thank you so much. It's such a relief," she sighed.

"Janice, anything I can do to make your lot easier gives me satisfaction too. We're a team," I reminded her gently, "we have to work together. Let me know after you've seen Doctor Kenny. Ask her to call me. Then we'll start treatment."

"Thanks, Doctor Berger." She sounded as a though she was smiling. "I have an appointment next week, so if all goes according to plan, I'll be able to begin treatment the following week."

It is impossible to keep track of all my patients except when they come for their appointments. If I do not specifically put a patient's chart in my desk, ask my secretaries (who keep a log book of messages), to remind me, or write myself a note, I have only my memory and a sixth sense to feel when a patient is overdue to see me. Sometimes I will realize that I have missed someone six months or even a year after they should have returned. I might read a journal article, hear a lecture, or just "pop," the thought of a patient comes into my head. I will then usually investigate and find out why there was no follow-up. Frequently the patient has forgotten, if it were only a routine follow-up. Sometimes the patient might have seen one of my partners if I were away. Some may have gone to Florida for the winter and feel comfortable enough to postpone examinations. Occasionally, the patient or family may have decided to seek care elsewhere or just stop treatment or follow-up exams and didn't inform me.

With Janice and her case, I was attached in a more-than-average-patient way and so put her chart in my top left-hand drawer. The space was limited to a select number of charts, but it helped me remember to check Janice's chart weekly.

Although I knew if she went to Memorial Sloan-Kettering for her routine care it would be much easier on me emotionally, I didn't want to lose her for two reasons. It was unlikely that Memorial would have any magical or more effective treatment than we had out on the Island, and also it would be very taxing to have her travel the sixty miles into Manhattan for regular treatments.

A few patients have elected to do this. Usually it is because there are certain investigational regimens requiring treatment only at Memorial or the physicians there may be able to administer drugs which are not yet approved for general use by the Food and Drug Administration and therefore not available to us. Some patients just go because they feel more comfortable being in a renowned cancer treatment center. Generally, because most of the physicians at Memorial have been confident in our care and expertise, the collaboration between Memorial and our group has worked well for many of our mutual patients.

Three weeks after my last conversation with Janice, my office nurses informed me while I was in with a patient that Doctor Natale Kenny was on the phone. Anticipating that her call was probably about Janice, I politely and quickly excused myself from the patient's room and went to my office to pick up the receiver.

"Hi, Natale! How are things in the snake pit?" I asked.

"The usual craziness," she replied succinctly although I well understood all she implied. "Listen. I just saw this young woman, Janice Bond. I offered her a choice of our new investigational chemotherapy protocol. When I told her it was not proven to be any more effective than standard therapy, she insisted on going back to you."

"I'm not surprised. What would you give her if you were in my shoes?"

"5-FU [fluorouracil], combined with leucovorin and CPT-11," Natale shot back. We continued by discussing doses and scheduling. After we agreed on a relatively standard regimen, Natale explained, "I'm sending her for an abdominal CAT scan since she never had one. You can use it as a baseline. She'll be getting it out by you. I'll be here if she wants to come back along the way, but she really likes and trusts you."

"I put myself on the line for this lady, and I have a feeling this is going to be tough on everybody," I replied candidly.

"Nice woman, but very angry." Natale confided. "She's going to require a lot of time—which is something I don't have so I think she made the right choice sticking with you."

"Well, thanks for seeing her and getting back to me."

"I'll send you a note about what we just discussed as soon as I can," Natale hesitated and then sighed, "but I'm already behind on my dictation, so don't look for it too soon."

"Don't worry," I assured her, "I've been there. As long as everyone knows what's happening. Speak to you soon. Bye."

I went back to work that day, expecting to see Janice in my office for a follow-up appointment sometime within the next week. Instead, a week later, I received a call from Janice's sister who lived in New Jersey.

"Hello," I said over the intercom from my desk as I began shuffling through all the unread reports for Janice Bond's abdominal CAT scan. "How can I help you?"

"Hello, Doctor Berger, my name is Gina Gables, Janice's sister. I called to speak with you about her condition," she said matter-of-factly. "Have you spoken to Doctor Kenny at Sloan-Kettering yet?"

"Yes, last week."

"Then you know that Janice wants to return to your care as soon as possible, although Doctor Kenny sent her for a CAT scan last Thursday?"

"Yes." I said to assure her. "I agreed with Doctor Kenny about the CAT scan."

"Do you have the report back yet?" Gina asked pointedly.

"Yes, I have it right here," having just found it. "I'm afraid I don't have good news for you." Reading directly to her from the report, "there is extensive metastatic disease to both lobes of the liver and peritoneal cavity," I said dejectedly.

There was silence for a few seconds, then a huskier, choked voice retorted, "Does Janice know this?" Gina was horrified.

I sensed I was walking on thin ice with Janice's sister, as I replied. "When I first saw Janice in the hospital, I explained that the disease was spread beyond the primary lesion which was removed from the colon. She didn't ask exactly where it was, and I didn't hammer her over the head with it at the time. Some patients don't want to know all the details right away, but gradually get around to asking. A few need all the information immediately and some never want to know. To me, Janice seemed overwhelmed by knowing she wasn't surgically resectable so there was no need to go further at the time."

"Well you're right about that!" Gina responded, "But if she finds out it's in her liver, she'll quit fighting. Please, Doctor Berger," Gina pleaded before her voice broke off, "don't tell her..."

"I understand how you feel, but promising to not tell her puts me in a difficult position. What should I tell her when she asks for the results of her CAT scan?" I countered.

"Tell her something hopeful... I don't know what... tell her there was something minor you have to treat her for."

"I usually try to keep the patients' spirits up, but my relationship with your sister needs to be based on mutual trust, which requires the truth. I'll do everything possible to paint as hopeful a picture as I can for Janice, but you must understand that short of a miracle, her outlook is very poor. This awful truth needs to be dealt with on many levels. I'll begin by giving her a small amount of the truth at a time, but if she pushes me to know more, I will tell her everything."

"I'm sorry. I don't mean to say you should lie to her, but I love my sister so much, I'm just so afraid for her. I don't want to see her in pain or suffering. But worse, I don't want her to give up and not fight..." Gina had managed to hold in her sobs until her final words, then they overwhelmed her so she could no longer speak.

"Please understand. By keeping her in the dark, I'm afraid I would be doing her a disservice." I felt deeply uncomfortable. I very much would have liked to tell Gina and Janice that all would be well.

In fact, things might go well for a while but inevitably it would be my responsibility and painful burden to take Janice, her sister, and the rest of the family together on Janice's last voyage. I could only hope it would not be too stressful physically or psychologically for any of us. I knew, however, that like birth there was often a long and arduous labor involved in death. Certainly, this end to the life cycle was not the joyous occasion of birth. At best, I hoped that I could help in prolonging Janice's life and then help her achieve a peaceful and pain-free death. She would need to work hard at this as well. The work was not going to come easily for either of us.

"I'm sure you'll do the right thing, Doctor Berger," Gina said rather curtly. I wondered if she were just upset about the news, or if she were angry because she could not sway me entirely to her way of thinking, or if she was incapable of speaking because she was crying.

"If you want to speak with me again, please don't hesitate to call." I sympathized with her dilemma, I could only hope she understood mine.

"Thank you very much, 'bye!'"

My mind went back to Janice, but the intercom came to life and my secretary said there were three patients in rooms ready to be seen. As I left my office, I thought: And each one of them has an entire story complete with family, fears, and other emotions that had to be dealt with. What an impossible task, I thought as I moved on to do the best that I could.

Janice came in by herself the next week. I walked into the examining room to see a poised and seemingly self-confident woman standing before me. "Hi, Doctor Berger. Well, I'm back and ready to beat this thing!" she exclaimed.

"It's nice to see you, Janice," I said calmly. "How was your appointment at Memorial?"

"Awful! It's a factory there and everyone is a number."

"I spoke with Doctor Kenny about you. She seemed very concerned."

"Well. She was okay, but I feel much more comfortable with you. Did she tell you that she ordered the CAT scan for me?"

"Yes. The CAT scan was ordered as a baseline to see how the treatment will affect your disease. We also came to an agreement on your treatment regimen."

"How long will it take until I'm cured?" she asked.

"Wait a minute, we haven't even started, and you're already asking when it's finished." Her unrealistic optimism was going to make it very hard to tell her the truth.

"Just tell me what I have to do and I'll do it. I want to get this over with so I can get on with my life," she stated defiantly.

"All right, let's talk about the treatment." I was relieved to be able to sidetrack her from the results of the CAT scan. "Leucovorin, 5-fluorouracil, and CPT-11 are the drugs we'll be using. The first is a vitamin which increases the effectiveness of the second which is a chemotherapeutic agent. The third is also an effective drug that helps kill your cancer cells. The three drugs are given weekly by injection into the vein. The entire treatment only takes about two hours. Each time you come we'll check a blood count to be sure its high enough to administer the therapy to you."

"What about side effects?" she countered.

"I was getting to that next. Like all chemo, side effects vary from patient to patient with some experiencing nearly no toxicity to others who require hospitalization for severe side effects. With your particular therapy there have been a few patients who have gotten nauseated from them, though vomiting is rare. Hair loss is minimal to nil. The most common side effects are mild mouth sores, diarrhea, and an occasional bloody nose. We adjust the doses of the drugs each week according to your blood count and side effects. The white blood cell count can be affected and if it goes too low, you will be more susceptible to infections than you would be otherwise. If you feel poorly—really out of the ordinary or having shaking chills or a fever over 101, you should call us. Please don't wait if you have any of these symptoms. Also any bleeding that is not stopped should be reported since the chemotherapy can also lower your platelets and platelets are the cells that help stop bleeding. Please don't take any aspirin or aspirin-containing drugs since they cripple the function of the platelets you do have. Tylenol is okay to use. Also you can drink alcohol in modest amounts." I threw the last idea in to cover all the bases as it was a frequently asked question. "Do you have any questions?"

"A few. When do we begin?"

"Right now," I said, "if you're ready."

"When will you know if it's working and how do you tell?"

"It will take about a month or so 'til I know if we are affecting the tumor cells. I can tell by examining you and doing blood tests and after about three or four months repeating a CAT scan." I just realized that my reference to her CAT scan was going to lead back to dealing with the results.

"What did my CAT scan show?" she asked predictably.

"Your CAT scan confirmed what we already knew... that the disease has spread beyond the colon and that there are cells there that need to be treated," I tried to say this all matter-of-factly. Time was beginning to put its pressure on me. While I would've liked to continue our conversation at a leisurely pace, giving Janice all the time she needed to ask her questions and receive the answers, I knew that my other patients did not like being kept waiting and would rightfully complain if they felt their delay was unreasonable.

"How many tumors are there?" she asked after some hesitation.

I wasn't sure she really wanted the answer or was ready for it, but fortunately I could not give it to her. "The radiologist did not give a number in his report."

"Didn't you see the scan yourself?" she challenged.

"No. I rely on the radiologist who has a much more highly trained eye than mine to read it. It's a baseline study. I'll look at it when we do the next one so I can judge the degree of response that we get," I said hopefully.

Janice seemed satisfied to end our conversation on a positive note. "Okay, let's start."

On two counts, I was relieved that I had avoided the issue of her liver involvement. First, I thought she would be devastated by the information as her sister had warned me, and, secondly, I did not have the time to go into a longer and more emotional conversation with her about it just then. I really felt I had to go on to the next patient.

"Okay, Janice. Ruth, our head nurse, will be in shortly with your treatment. I'll see you next week. Call if you have any problems."

"Thanks for all your time," she called after me as I left the room.

"No problem," I shouted back over my shoulder.

Over the next several weeks, I saw Janice as she came in for her regular therapy. The chemo did not cause any nausea, vomiting, or mouth sores but did cause an increased looseness of her bowel movement through her colostomy.

Until her surgery Janice had been a particularly fastidious woman who always strove to look her best. In order to hide the colostomy and appliance, Janice had to change her style of dress from tight, body-conforming clothing to loose-hanging outfits. She absolutely hated having to deal with this change in lifestyle because it was a constant reminder of her cancer and its total impact on her life.

"When can I get rid of this wretched thing?" she complained during one visit.

"Have you asked your surgeon?" I countered.

"He said it's up to you."

"Well," I said slowly stalling for time, "let's see how you do with the chemo and then we'll cross that bridge."

There was no sense putting Janice through a second surgical procedure if the treatments were not working. Furthermore, I knew there were tumors in her pelvis, which when they began growing might obstruct her rectum and sigmoid colon, making it necessary to re-operate and do another diverting colostomy.

"When will we know how I've been doing with chemotherapy?" she asked.

"I'm drawing a CEA blood test today, a test we'll do about once a month, although we won't have the results 'til next week. It gives us an idea on a monthly basis in which direction we're headed," I said.

"I don't understand. What's a CEA?"

"The CEA stands for Carcino Embryonic Antigen. It's a blood test that corresponds to the number of cancer cells that secrete it in your body. Generally, if the cancer cells are being killed, the CEA should go down. If the chemotherapy is working, this will be confirmed as well. If the CEA is increasing we will need to do another CAT scan to confirm that the tumor cells are increasing before we change our treatment."

"It's working," she said angrily. "I'm beating this damned thing! I have two children who depend on me and who I love more than anything in this world and who I have to see grow up and get married."

My stomach was in a knot. I knew Janice's kids were teenagers, a boy of fourteen and a daughter of sixteen, similar in ages to my own children. The chance of her getting her wish, short of a miracle, was near zero. What could I say? How was I to respond without being elusive or untruthful? I wanted to be both honest and hopeful with Janice, even though she was not yet ready to hear that my "wretched" treatments could only prolong her life by months; we'd be very lucky if she could get years more. I knew I had to prepare her for this, but it had to be done slowly and carefully. It had to be done based upon her questions and her readiness to handle it.

It was also very unpleasant for me. I had done it so many times before, yet each year it seemed only to be getting harder for me personally.

How could I make peace with this dying business so it wasn't so anxiety-provoking and so stressful? How did my patients do it? I realized that if I could somehow feel less anxious about my own demise and separation from life, I might be able to impart more comfort and peace to my patients. After years of denying any thoughts of an unscientific nature about the afterlife or the soul, my life's work was forcing me to confront that very issue.

"My bet, Janice, is that it's working," I said skimming over all these thoughts instantly. Then I added with emphasis, "We'll know more next week." I had spoken the truth. If treatment was not doing something, her disease would be obviously more progressed than it was.

"So let's get this over with and we have a date for next week," she said.

I left the room hoping that the following week would bring news of a response. I knew that eventually the painful conversations I dreaded would need to take place. I just couldn't handle it at this particular time.

After all, that was the business I was in—the time business. Patients put themselves under my care with the hope that my knowledge and skill along with luck or divine guidance would prolong their time on earth.

The week went quickly by, and before I knew it, Janice was again in front of me, appearing quite anxious.

"Okay! How am I doing?" she demanded.

I opened her chart and saw that her last week's CEA had dropped significantly, enough for me to shout triumphantly with sincere and great relief, "We're winning! The tumors are regressing. It looks like we killed more than half of them!"

Tears of relief and joy were running down her cheeks. I went over to her and soothed her, gently caressing her back. Her tears gave way to sobs, then she got up and hugged me tightly, nuzzling her head into my shoulder. I held her in my arms, glad to be able to be a part of her catharsis. All the pent-up anger, fears, and tension were finding a healthy release. At the same time, I knew that this moment of joyous victory was temporary, and this made me feel uncomfortable. The course of events were sure to turn for Janice, yet she was still deluding herself into believing that she would be cured or at least survive for many years.

How could I give her a realistic picture of the future without making her lose hope? Was it for her benefit or mine that I was keeping silent? If I told her the full truth would I avert the inevitable anger that is often times aimed at me for holding out too long? If I told her now, she could not accuse me of deceiving her; yet if I didn't, I risked even greater anger when the disease ultimately progressed. What was the best tack? It differed for each patient. How difficult it was to get a bead on each individual in so short a time and make decisions that would have such a strong impact on their lives. The feeling was one of tremendous power and at the same time great fear and anxiety. It seemed so difficult to get this right, to do the right thing for each patient, to say it the right way, to do it without destroying all hope, yet without lying, so that when the chips were really down, and pain and suffering needed to be allayed, the patient would believe me when I said I would and could keep them comfortable in the end.

Janice lifted her head off my shoulder, wiped her eyes, looked directly at me, and asked, "How long will it take to kill the rest of the tumor cells if half of them are dead already with treatment?"

There would be no emotional respite today. I replied, "The problem with these treatments is that the tumor cells usually figure out ways to resist the treatments before all the cells are killed." There I said it as tactfully as possible. Anyone, not using massive denial, would figure out the rest.

"Then what do you do?" she challenged.

"We change the drugs to ones the cells haven't seen before," I countered.

This was like Ping-Pong. I hoped that her side would retire and not make me continue volleying the ball.

But she came back. "Does it always work?"

"Unfortunately, no. The cells somehow develop what we call multi-drug resistance. The cells are able to pump the drug out before it can do any major damage."

Janice began to look more apprehensive now. Finally, the truth had sunk in. One moment, she was experiencing tears of joy, the next, as I feared, tears of anger and hopelessness, all a result of my interpretation and statements about what I felt would be the course of events. Who was I to be able to predict Janice's future? I could have played it safe and just have said, "We'll do our best and see what happens." But that seemed like a cop-out to me. She was asking the questions which I felt required responsible answers given in the best way I could.

This was my style, always being fine-tuned and changed as I learned to navigate the treacherous seas of physician-patient interaction.

"Are you telling me that I won't be cured?" she questioned bluntly.

"There is continuous research ongoing all over the world. If new drugs are developed that can kill resistant cells, we will be able to make further progress with your tumor cells..." Again, I left the door of hope open. I, however, was pessimistic. No drugs had yet been discovered to cure metastatic colonic carcinoma since man was placed on this earth. It seemed hardly possible for it to happen before Janice's cancer grew to the point of terminating her life.

"It's not fair," she yelled angrily. "My kids need me! I'm not ready to die."

"Janice, I just gave you very good news. We're winning for now. Concentrate on that. Enjoy today. Enjoy tomorrow. Be with your children. No one knows how many tomorrows we have."

I wanted and needed to end this conversation. I had to move on. I was emotionally drained. I hoped I would have a respite with the next few patients. I didn't know how much reserve I had. I might need to do this several more times today. I understood once again why many doctors need to stop feeling—the pain is too great. I knew, however, that if I stopped allowing myself to feel the pain, I would be much less effective at allaying it, and then neither I nor the patient would get much from the relationship.

Janice didn't want to go further that day. She received her chemotherapy and left, but my thoughts returned to her many times that day and week.

Fortunately Janice continued to respond to therapy week after week. There were visits which went easily and quickly. These tended to occur when both she and I were in a hurry or when she was emotionally and physically stable. Other visits were not so smooth. They were reiteration sessions, in which Janice would rehash questions and answers, hoping each time to hear answers different from the ones I had previously given her. This was undoubtedly part of the bargaining process. We were dealing with the seesaw of being able, then unable, to cope with the diagnosis, prognosis, and change in her lifestyle that was foisted upon her by her disease.

Once a month, I would spend longer periods of time with her, doing a full physical examination, discussing the results of her blood tests including the progress of the CEA which had been drawn the week before. During this time, we had fallen into a reasonably comfortable pattern, even with the sword of Damocles hanging over our heads. I knew her questions and she knew my answers.

Yet, I knew there would be the inevitable rough waters to navigate ahead. Each time her blood tests were drawn there was a tension which could only be dissipated by results indicating continued regression of her tumors.

"How is my CEA?" she asked nervously on one such visit.

"It's continuing its downward trend," I replied thankful that yet another month of responsiveness to chemotherapy was under our belts, and relieved that I did not have to deal with the emotional upheaval that awaited us on the day when the progression of the disease became evident.

"How much longer do you think this will work?" she queried.

"Janice, I don't know if it's weeks or months," I left years out deliberately. "The important thing to me and more importantly to you is to appreciate each day. That's all any of us have, yet we are all so intent on the future that we have trouble appreciating today."

"That's easy for you to say," she grumbled, "since your tomorrows are not in doubt right now. Mine are and I can't help thinking about them."

"Believe me, I understand, but concerns about your future have to be put in perspective. If obsession with the future takes over, the 'now' will mean little to you. Look, your disease is still responding. Go home to your family. Be with them. Hug them, love them. Do the things you all like to do together. Try to enjoy the moments. We'll do all we can to secure as much of the future for you as possible. That's all we can do. Go with the flow a little and try to let go a little too." These were the difficult lessons I had begun learning as I watched others die. They may have sounded like platitudes to some, but they were fast developing into my own philosophy of dealing with death and separation. I was beginning to see how significantly, if embraced and understood, they could change a person's life.

"Easy for you to say." Janice rebutted. Apparently, she wasn't ready to develop such a philosophy to help her through her ordeal.

Exasperated that these important messages had fallen upon deaf ears, I finally said, "See you next week."

"Thanks for everything," she said warmly as if all the dueling hadn't taken place.

"You're very welcome," I said leaving the exam room.

Many more months did pass uneventfully, enabling Janice to enjoy her well-time with her family. However, one day she wore a look of pain and fear on her face as I walked into the examining room.

"What's the matter?" I asked immediately.

"I have a pain in between my shoulder blades. It's a dull ache that began about two weeks ago, I hardly noticed it as first, I thought it was muscle strain or something. I realize it's getting worse. Now it's so bad it's been keeping me up at night." Her voice shook apprehensively.

Masking my true concern so as not to increase her anxiety, I calmly asked several other questions and concluded, "Well, I think we ought to get a bone scan and repeat your blood tests." I knew in my gut what was happening and so did Janice, but neither of us was ready to acknowledge it to the other. She wasn't

157

ready to hear me say it, and I wasn't emotionally prepared to face the onslaught of emotion that speaking it openly would certainly incur.

"Okay, Doctor Berger. If you think it's necessary. When should I get the bone scan?" She fidgeted but appeared calmed by my matter-of-fact delivery.

"This week." I said while mentally checking my next week schedule, ensuring that I would be in the office and able to deal with the inevitable bad news that would be engendered by the results.

"Okay," was all Janice said.

"I'll have the nurses come in and get your blood so we'll have all the results ready for your next week's visit." I said all this impassively as I extricated myself from the room. Out of the corner of my eyes, I notice one of Janice's hands rise up to wipe a tear from her eye. I wanted to go back in and tell her everything was going to be all right. I wanted to tell her not to worry. I wanted to make her pain be something other than what it was, but I couldn't. I continued to the nurses' station and wrote out the request slip for the bone scan and a prescription for a narcotic pain medication, oxycodone. I wiped a tear from the corner of my eye.

During the week that followed, I performed my usual hospital rounds, attended conferences or evening meetings, and occasionally thought about Janice and our upcoming appointment. It was not, I knew, going to be the comfortable routine we had in the past. We were passing another milestone in the course of her disease.

Janice's scheduled appointment arrived and I was suddenly faced with going into the room. To prepare myself, I opened the chart and looked at the laboratory results. Her CEA had increased substantially, which warned me that the bone scan would be positive. I was surprised, however, by the extensiveness of the disease in her back since she had only complained about one area. I was now resigned to the progression of Janice's disease and very disappointed at the extent and rapidity with which it had become resistant to therapy.

"Well, am I cured yet, Doc?" she jibed sarcastically.

"Almost..." I fumbled, not thinking quickly enough for a more appropriate retort.

"My pain is worse than last week—even with your pain medicine," she taunted.

"I'm sorry, Janice," I felt awful. "It's understandable why you have this severe pain in your back. Your bone scan shows the disease has spread there. Your blood tests confirm it."

"I just can't believe this is happening to me..." she said softly. Something in her voice implied, despite her words, that perhaps she had begun to accept it. Perhaps the week between visits, living with the worsening pain and knowing she needed a reevaluation which included a bone scan, was enough for the reality of her situation to sink in, or was it my imagination?

"There are a number of things we can do to alleviate your pain and try to get your disease back under control," Still I offered hope, but Janice was not listening. Instead, she was sobbing quietly. As I walked over to her to place a calming hand on her shoulder, she allowed her sadness and frustration to vent with long, deep sobs and streams of tears. My eyes welled up as I stood quietly above her.

Numerous and fleeting questions occurred to me as we stood there together. What was the purpose or meaning of all this suffering? Why didn't we—the medical profession, the human race—have the right answers to rid ourselves of incurable diseases like Janice's? Why did I go into this field of such sadness and such frustration? No answers were forthcoming. The sobs turned to quiet whimpers and soon silence.

Breaking the silence, I explained the radiation therapy that would most likely alleviate Janice's pain. We discussed new chemo treatments, drugs her tumor cells had not yet seen. A new hope began to take root in Janice; it was her survival instinct that was as basic to the human spirit as life itself. She left the office with a new attitude, ready to do further battle with her disease. I admired her courage and determination.

I half expected it when Janice's sister, Gina, called later that day. Even though my schedule had lagged, I felt I had to take the call, rather than postpone answering until office visits had ended. If I could bring her sister up to date (although I figured Gina grasped the entire picture clearly), perhaps Gina could give Janice immediate solace and help her muster the strength to continue.

One thing was certain. I was going to have to keep our conversation brief. "Hello, Gina. I'm sure Janice has told you about the setbacks we're having. Do you have any questions?"

"She told me the cancer has spread to the area between her shoulder blades. Is it anywhere else?"

"Unfortunately, it appears to be in most of the vertebral bodies of her back. Janice wasn't ready to ask that question, so she's not aware of the extensiveness of her disease. I think it's better not even to mention it at this point."

There was an uncomfortable silence. A more muted voice replied, "Of course, of course."

"We will be starting radiation treatments right away. I believe they will relieve her pain to a great degree and next week we'll change her chemo treatments. Our hope is to find new drugs that will affect her tumor cells."

"How long will it take before you know if it's working?" Gina asked.

"About four to six weeks. Maybe a little more."

"My sister trusts you, and so do I. I know you are doing everything you can for her."

"I'm very fond of her and will continue to do whatever can be done."

"Thank you. Can I call you again to see how she is doing?"

"Any time."

Gina's voice began to tremble as she thanked me again and said good-bye.

Janice's radiation began working relatively quickly and her spirits improved. She tolerated very well the new, weekly chemotherapy, but after four weeks she began complaining of new pains in her back, this time lower.

I knew this was a very bad sign. I asked the radiotherapist who was treating her upper back to reevaluate her for radiation therapy. Clearly the new area of pain was involved with the cancer on her bone scan. I knew I was running out of drugs if the new ones weren't working. My fears were confirmed when her CEA increased, demonstrating that her new pains were in fact a progression of her disease. The chemo wasn't working. Now what?

"Janice Bond is on the phone, in severe pain. She says she can't stand it anymore," my secretary Sue said sympathetically.

"Tell her to come right in," I replied.

An hour later Janice was in an examining room with a close friend who had accompanied her to my office on many occasions. One look at Janice and it was clear that the quickest way, perhaps the only option, was to give her the kind of pain control she needed by hospitalizing her.

"Doctor Berger," she began hoarsely as if speaking hurt too much, "I can't deal with this pain anymore. Please do something," she entreated.

"I can give you the best and quickest pain relief in the hospital, Janice. We'll start a drip of morphine sulfate, M.S., that can be raised quickly until you are out of pain."

"Anything, please. Is there anything I can get now?"

"Of course." I came back with twenty mg. of morphine, injecting it slowly and directly into her vein. I was not timid about using effective doses of narcotics to get my patients out of pain. Over time, I had grown comfortable with the doses and indications for this use of morphine.

"Does that take the edge off?" I questioned after a minute or two.

Janice seemed less tense and replied, "It's a little better."

"Stay still on the table while I write up your orders and arrange for your bed at the hospital. I'll be back in a short while."

With orders in hand and bed arranged, I returned to the room. Janice was still in pain. I injected another ten mg. of morphine which enabled us to get her up and into her friend's car, and then she was brought to the Oncology Unit at the hospital. The nurses there called me to verify my orders. After answering some questions, I recommended that they quickly get the M.S. drip started, which they assured me they were doing. As I hung up, I admitted to thinking that Janice would not leave the hospital again.

I did not get to see Janice for a few days, due to the way our group is scheduled to round at the hospitals. She was told she would be seen by a different member of our group each day so that she would not feel abandoned by me. Nevertheless, it is often quite difficult when a patient has come to trust and rely upon only one physician for care, consolation or hope, and that physician appears to come up short during a very needy period.

In order to compensate for this apparent shortcoming on my part, I proceeded into Janice's room with a warm smile.

"Hi, Janice," I hailed from the doorway. I stepped into her room and went over to her bedside. "How are you doing?" She appeared much more comfortable than when I last saw her.

"Terrible," she retorted. "I have to get out of here, I miss my family."

Ignoring her initial words for the moment, but at the same time probing to see if her demand was possible, I asked, "Is your pain under control?"

"It's better but not gone. When they increase the morphine so the pain goes away, I get confused and upset so they turn it down."

"Do you prefer to have a little pain and be more lucid?"

"Yes. I hate when I lose control," she said with a grimace.

"Okay, maybe I can add a medication which will increase the effect of the morphine without increasing your confusion. I think it's worth trying. You have to understand that we have to do this by trial and error. If it works, great. If it increases your confusion, I'll instruct the nurses to stop it."

"Whatever you think is right, I'll go along with, but when do you think I can go home?" she reiterated her question.

"As soon as you get your pain under control and get a care system into your home so that we know you'll be comfortable and well taken care of."

She nodded as though she understood and then pleaded. "Please sit down, Doctor Berger. I'm scared and upset. I don't know how to deal with all this. I'm afraid of dying. Nobody has ever discussed dying. I've suffered enough. I want the pain and suffering to be over, but I'm not ready to leave my kids." She began to cry. I reached out to hold her hand. I recalled that I had done so the first time I met her about a year ago. As I held her hand now, I knew that this moment of final resignation, solace, and comfort had been predictable from the start. The roller-coaster ride was near its end. What was not predictable was the experiences in between—the high climbs to victory, the sudden frightening drops into despair and defeat. The spectrum of human emotions and variability in dealing with disease is tremendous and each individual brings his or her own specialness in dealing with this end of the life cycle.

"Janice," I said softly. "From what I know of you, you have been a loving and giving mother. Your children have gotten a lot of strength and other good qualities from you. That will live on in them. My own views of death have evolved over time. Maybe my sharing them with you might help."

"I... I'd like to hear them," she said somewhat choked by her tears.

I remembered back to the previous summer when I decided to explore my own fears of my mortality. Faced with death every day of my professional life, I decided I could not ignore my feelings any longer. I began to voraciously read as many books as I could about death, reincarnation, religion, and meditation. My anxieties about eternal separation from all I knew and loved was far greater than I supposed. It only became clear to me as I found myself better able to sit with my patients and their families and explore these issues with greater equanimity than ever before. My views evolved, and in fact, are still evolving as I continue to interact and learn from the myriad of patients that are going through the final phases of life as Janice now was.

"Janice, you will not die."

"I won't?" she said with incredulity in her voice.

"Well, what I mean by that is, your body will die, but the inner essence of you never dies."

"How do you know that?" she asked with a calmer demeanor.

"I cannot prove this to you, Janice, but there is too much data and indirect evidence to make me believe that your life just ends and that's it. For example, we know that water is something you can touch and feel. It can change from ice to water. Both of these are visible and palpable. Yet when water changes into steam, the vapor emanates into the atmosphere and can no longer be seen or felt. Does that mean the water molecules are not present? Of course not. Matter is neither created nor destroyed. I believe that this is true with the human spirit. A human soul can never be destroyed. Your essence will never disappear."

Her grip on my hand tightened and she gave such a sigh of relief so great that I could almost see the tension lift away from her. She uttered a quiet, "Thank you," and I left.

The next two weeks were rocky ones for Janice and her family. Her condition continued to deteriorate. Her morphine and tranquilizer needed to be increased to keep her comfortable and calm. She was able to say her good-byes and straightened out her affairs with her family.

One morning about two weeks later, Janice quietly passed on to the next phase.

10~

And Tomorrow's Miracles?

The thank-you card had the happy couple's wedding picture on the front. Lydia looked wonderful. The photo had captured her beautiful face smiling at her handsome groom as they locked arms and bowed gracefully forward to enter the limousine. That moment was frozen and now reduced to a wallet size photograph resting in the palm of my hand.

The sentiment and emotions associated with this card were by no means reduced for me. Lydia was a triumph, my triumph, and although I hadn't attended the wedding, I felt like a member of her family and was deeply moved by paternal joy as I studied the picture.

It was ten years ago when she cried about undergoing further treatment for her ovarian cancer. She was nineteen, scared, and in a life-threatening situation, yet she resumed treatment as long as I continued to coax, cajole, joke, and bolster her spirits. Although it was emotionally draining for me, I tried to give as much as possible to help Lydia and her loving family through the worst of it. Within a few short months, we all saw a great response. Finally, when treatment ended for her, she remained disease-free for three full years. Now, six years later, she was alive, well, just married, and officially cured.

Opening the thank-you card, I read the message scrawled inside:

Our wedding day would not have been possible, and I wouldn't have found the woman who is now my wife if it weren't for your help years ago.

Lydia's family has never forgotten your compassionate efforts to fight her cancer, and because they have often talked about you and praised you for giving them and Lydia a second chance at life, I feel that I should add my voice of gratitude to theirs.

During this joyful time, Lydia and I toast you, your compassion, dedication, expertise, and most of all your successes: past, present, and future, with deepest, heartfelt thanks.

163

It was signed by both Lydia and Anthony Gray, Lydia's new husband.

Even though the success itself was very sweet, this tangible appreciation fortified me. It affected me personally: I loved it and needed it. Getting people through difficult times with finesse, helping them through terror, anxiety, and frustration, was a requisite of the job, but also it was something I always strove to do to the best of my ability. Such appreciation's from my patients were trophies of my achievements.

Like every other human being in the world, doctors respond well to positive reinforcement. Often the reward is love, a deep bond between doctor and patient based on trust, caring, and appreciation. When a doctor experiences this with a patient, undoubtedly it makes the work worthwhile. It's what practicing medicine should be all about.

I have felt this appreciation throughout the years: when Janice Bond's family recognized what I had done to try to prolong her life and give her quality time, when Jeanne Murray was able to live long enough to see her youngest daughter's wedding and her newest grandchild; when Lynn Moline was cured of her aggressive NonHodgkins Lymphoma and was even able to have children; when Joe Fry, once so close to death, was able to be cured of his germ cell testicular cancer that had metastasized to his lungs and brain. Every patient and family member who valued my efforts, whether the outcome was cure, prolonged remission, or even an unexpectedly good response, renewed my sense of accomplishment. Even when the patient was at the end-stage of disease, if the survivors recognized my attempts to palliate the pain, reduce the suffering, and ease the patient to a peaceful death that the family members could accept, then somehow I was given the strength to go on to the next patient in need.

As I put the thank-you card down on the desk, I heard a loud rap on my office door. Even though I was on my lunch break, I was used to these interruptions. "Come in," I called.

Although I had expected to see the familiar face of one of my staff, instead Ellen Langston, a long-time patient, popped her head hesitantly through the door and then smiled when she saw me. "I have something to show you..." she said and then stepped into the room. In her arms was her newborn baby girl, now six weeks old and as delicate as a tiny doll.

"Marissa, meet Doctor Berger," she cooed to her baby and lifted her daughter's curled fist in an enthusiastic wave.

"Ellen, she's beautiful!" I rose from my desk and went over to get a closer look.

At twenty-eight years of age, Ellen was an extremely tall, thin, attractive young woman. I had treated her about six years earlier for a very bulky Hodgkins disease. She had been staged elsewhere and had begun radiation treatment, when she came to me seeking different therapy.

Despite great personal difficulties in her life, Ellen was a survivor. When she was first diagnosed, her home life was in turmoil, her boyfriend was unsupportive, and her rapidly consuming illness was making it difficult for her to keep up with her job as a hairdresser. Yet, she had the presence of mind to sense that the radiation treatment which had been prescribed for her wasn't effective against her disease. When she bluntly told her previous doctors that she wasn't pleased with the radiation—she didn't believe it was working—they suggested she see me.

Her instincts were right. I informed her that, in my opinion, a better way of treating her was with an aggressive chemotherapy regimen. She agreed, and although the treatment was difficult for her, she responded very well, and soon all evidence of her cancer had vanished.

I was encouraged by this success and all the more confident in the protocol, even though Ellen began to fight me along the way. Chemo was harsh, some drugs having more toxic side effects than others, and she wasn't a passive person to just sit around and take it without loudly voicing her complaints.

Her personal life complicated things further. She and her boyfriend had broken up. He couldn't deal with the stress of her disease. All during her treatment, her parents fretted so much about her illness, as if she were at end-stage, that they weren't able to supply her with any support. This made it hard on her and harder for me to convince her that her progress was steady and quite good.

Toward the end, I had to plead with her to finish her treatment. Finally when it was done, I breathed a sigh of relief. Ellen had pulled through like a champ.

Her follow-up visits over a period of two years were without recurrence, although she became very thin. It scared me at first because I thought her Hodgkins may have been coming back.

"No!" Ellen snorted a laugh in reply to my question about her weight loss. "I did this on purpose. I couldn't stand the excess baggage. Besides," she added with a twisted smile, "I'm getting married in two months. I don't want to be a fat bride." Her relationship with a new fellow she had met in the last year had apparently worked out fine.

"Congratulations," I chuckled, relieved.

Less than a year ago, in her fifth year off any treatment, Ellen had come in for her routine checkup. "Guess what?" she asked, grinning at me while she sat on the examining table. "I'm pregnant!"

"That's wonderful!" I smiled back feeling genuinely happy at her good fortune. "How do you feel?"

"The truth?" she narrowed her eyes as if she was sizing me up for her answer. "Awful. Sick as a dog. Morning sickness stinks. It's lasting almost all day. But, nothing is as bad as your chemo." She pointed an accusing finger at me, "I'd rather have morning sickness."

I didn't mind the disparaging remark about my chemo. I was used to it.

No patient ever liked chemo, although some patients became psychologically dependent on the idea of treatment and would even become frightened if they were taken off it. Often, the end of treatment meant prolonged remission or the desirable cure, but other times it meant that the particular regimen was no longer worth the toxicity against a very resistant tumor. Fear then in such patients was justifiable. This was when I have had to find a balance, giving them the idea that there was still some treatment while finding something they could tolerate and which might provide them with at least some hope of stopping the inexorable progression of the disease. Ultimately as the patients started getting sicker, I would have to back off the doses. By then, because it had been a gradual process, these patients were better able to recognize and accept the inevitable.

"One other thing you ought to know," Ellen offered while I was examining her. "Donald and I are moving."

"Where?" I felt a tug. I hated losing touch with my successes.

"Not too far," she told me. "To Malverne. It's almost on the border between Queens and Nassau County."

"Oh." I was disappointed. Having people get on with their lives was something I wanted, although sometimes this meant that they moved on as well. At least this kind of separation—geographic in nature—was far more acceptable than the kind I frequently experienced with my patients. "What do you want to do about your future checkups?" I asked, mentally reviewing the list of oncologists in that area. "I know a few doctors…"

She stared menacingly at me for a moment. "Are you trying to get rid of me, Doc? I wouldn't dream of going to anybody else but you!" Then she laughed, "Really, it's no big deal to come out here for my annuals, especially 'cause I've still got friends around here that I'll be keeping in touch with. Besides, I wouldn't trust anybody else to check me. You cured my cancer," she said with emphasis.

This was the kind of lasting bond between patient and doctor I relished. It was because we both did something so positive—cured cancer—not just treated a cold.

Now, it was ten months after her move, but Ellen's appearance with her infant wasn't coinciding with her scheduled checkup. "I've just come to show off my kid!" she explained proudly as I stood next to her to see the baby. "Here!" Suddenly the infant was in my arms squirming awkwardly while Ellen was cooing happily in the cute baby's face. When the baby started to whimper, she took the bundle back and cradled her soothingly.

As I watched the two of them, I felt thrilled. Mother and child both had life because of my success with Ellen's treatment. It was one of the highest highs to know that someone who probably would have died had been cured because of my expertise and craft.   This was the gamble I took when I decided to go into oncology.

"I've got to run," she said, "we're on our way to visit the old neighborhood, but I thought you'd like to see my little darling."   Then she added thoughtfully, "For the first time in my whole life I feel like things have finally gotten good. And I want you to know that I appreciate what you did for me, too, even though I complained a lot while you were treating me.   Up to the time I got Hodgkin's, my life was a bitch!   Since then—when the damn chemotherapy ended—it's been great.   Again and again I think that if you hadn't helped me when you did, I would never have known life could be so, ya know, rewarding?   I really mean it, thanks."   She gave me a gentle peck on the cheek, then Ellen and baby were gone.

Her praise, added to the thank you note from Lydia and Anthony, made me beam.   Although my morning had been rough, their appreciation turned those disappointments around for me.

Doctors need appreciation maybe more so than the average person.   At least, that has been my impression over the years.   People who are in need of appreciation, love, and acceptance tend to go into medicine or health related fields.   Through helping others, the caregiver receives the rewards and fulfillment from being needed.   The more the giver feels needed and appreciated, the more the giver is willing to give.   When it comes right down to it, most health care professionals have a lot to give.

Sure there are hard-feeling, callused individuals who are in it just for the money, but there always are a few bad apples.   While such cases unfortunately have quite serious repercussions in medicine, these few should not be held representative of the greater number of dedicated people who in fact went into medicine because they needed to help other people.

The constant criticism about doctors is that they are unfeeling, uncaring practitioners.   Although they may sometimes appear this way, the humanity is there. They are people, under difficult circumstances, trying to do the best they can.   Yet, when patients' emotional and physical needs are so great, the human beings on the other side of the medical facades develop thicker skins.   This protective layer shields them from becoming too vulnerable and thus ineffective. It's a form of self-preservation, a normal human response.

I know I may have appeared that way to patients, while masking my own profound sympathy as I informed them of their circumstances.   Occasionally, in the past, my blunt honesty has made me seem callous and unfeeling to some patients.   Making an error in judgment, thinking those patients and family members were more ready to hear the stark truth, I have frightened them off.   So, they would go elsewhere, hopping from doctor to doctor until they found one who would tell them what they wanted to hear.   I would suffer the consequences of my less-than-gentle honesty by hearing about their anger through the feedback of nurses, my associates, referring physicians, and sometimes directly from the patients themselves. Usually I would feel badly about misrepresenting myself and my desire to help these lost patients. So, over the years, I have learned to temper my statements and test the waters better before giving prognostic decrees.

Different doctors have different approaches. A surgeon once told me, "You have to be very up front with these cancer patients." Maybe. My feeling through the years has been you can and should be positive for those patients who really have a chance at doing well.

However, personally, I would become nervous about being up and very positive in the face of someone who is going to do poorly. I don't like to bear the brunt of their anger if I had said they were going to do well and they didn't. They're often already angry at me because I'm the cancer doctor. I don't want to give them additional reason to be angry by leading them astray. I'd rather paint the more honest picture and then tell them there's lots of things we can still do to help. As I've often said before, I will not deprive them of hope.

When the patient's response is better than expected everyone rejoices. This is what happened for a sixty-eight year old man named Harry, who was the father-in-law of my personal friend Doctor Mark Frankel.

After losing both his parents to cancer, Mark was quite aware that Harry's odds at survival, once he was diagnosed with gastric cancer, wouldn't be good.

When I accepted the referral, I realized that Harry's case was only a bargain for time. The blood thinner he was on for his thrombophlebitis was making his gastric cancer bleed. The bleeding made it impossible to start him on chemotherapy even though his cancer was now metastatic to the liver. Radiation was also out of the question because it would cause further necrosis, more bleeding, and more problems, besides making him sicker still. As it stood, Harry would be lucky to have three months!

The solution came to me as if in a vision: a hepatic artery catheter with an infusion pump could be inserted during the required stomach surgery. Much like a pace maker, it could be buried under the skin of the abdominal cavity, and while the stomach was healing from the surgery, the catheter could pump chemo directly into the diseased liver. Unlike the usual intravenous method, the pump would treat the liver around-the-clock without toxic effects to the stomach or the rest of the body.

I became excited and quickly found a surgeon with great expertise in this delicate procedure. The catheter worked like a charm! After surgery, Harry's liver function tests were better, his tumor markers improved. Three months passed. To everyone's amazement, Harry was visibly better and stronger; even his appetite returned.

Another six months later, at the bar mitzvah of his grandson, Harry appeared remarkably well. Dressed in his fine tuxedo, he was nodding proudly, smiling warmly with his joyful wife, Martha, as the clear young voice of their grandson resounded melodically through the temple. Around Harry, other family members were exuberant on this glad occasion.

As a family friend and guest, I was seated five rows back which gave me a clear view of Harry. Looking at him, I was acutely aware that, while Harry had cheated death, the man's time was limited. It was somewhat miraculous that Harry was alive at all and able to attend the bar mitzvah. However, I was feeling gratified that my solution had an even better response than bargained for, giving Harry the opportunity to experience such an important milestone in his life—a milestone Harry would otherwise have missed. With that thought, I leaned back in my seat, folded my arms, and allowed tears of joy to fill my eyes as well. It was indeed a proud and happy moment for us all.

What made Harry's case so guilt-free for me was that I had laid my cards on the table. Since I stated his chances, as realistically as possible at the onset of treatment, his whole family was better able to appreciate his remarkable response. Unfortunately after the bar mitzvah, Harry's time ran out.

Weeks after, the constant pain from Harry's liver began to get worse. We attempted to control his pain, using various modalities which sometimes were effective and sometimes not. In time, it became obvious to him and his anxious wife that his response to second-line therapy wasn't lasting as long; his discomfort returned much more frequently. This meant that we would have to move to third-line treatment and hope it would control his pain as well as provide him with a reasonable amount of quality time.

By this point, each time the couple came to the office, Martha was in absolute terror of losing her husband, she loved him so much. Harry, on the other hand, looked weary and worn from treatment. Yet, instead of voicing his opinion, he would lean back in his seat, with arms folded, and let Martha ask all the questions. Indeed, when it came to what therapies were available, she was an informed woman. Throughout the early stages of Harry's treatment, she had investigated diligently the different methodologies for treating his cancer. Often she would grill me with questions about the latest reports.

I had done my investigative research, and read the medical journals on the most current treatments, so I had heard of nearly everything she might mention. She listened closely as I explained to her the reasons those other therapies weren't acceptable and although she trusted the accuracy of my knowledge on the subject, I could see her disappointment. She was desperate, more so than Harry, to fight for his life, and each of my informative answers was one less hope to hold.

"Martha," I had to bolster her spirits, "We are keeping on top of Harry's disease pretty much. I'm constantly checking through the literature to find any other protocols that might be more effective. I'm always consulting with other oncologists and radiologists about Harry's disease. Please believe me when I tell you that Harry has responded better than most." This was true. He had long passed the expected median survival for his disease.

"It's not enough," Martha replied in frustration, "I don't want to give up on any possibility. We're going to fight this all the way." Her determination was fierce.

Although I was worried that maybe Martha wasn't facing the fact that Harry had very little time left, I knew better than to be brutally honest about it. Instead, I stated as truthfully as possible, "No matter what happens, as long as you both want it, we'll continue to try the most effective treatments and hope for the best."

Ultimately, it was proven that the protocol I had chosen for Harry was indeed the best for him for a while, even though he still had pain. I felt none of the other treatments would be any more effective. Martha especially realized this and continued to have faith in me. However as treatment began to fail him, they insisted on more chemo and I was feeling uneasy.

Normally, for a terminal patient as old as Harry and in such depleted condition, the patient's quality of life was greatly compromised by the added toxicity of the drugs further down the line. Whereas it was Harry and Martha's choice to continue with treatment until the end, I wanted them to know that the time he had left, whether or not he received chemo, was minimal. As much as Harry seemed to understand, Martha willfully refused to acknowledge this. She would not give up on Harry. At their request, and after much consultation with my friend Mark and a team of specialists, I offered Harry and Martha one last antineoplastic therapy: radiation to the liver.

Harry had hardly begun radiation treatment when he deteriorated rapidly. His toes were darkening. Eventually they would become gangrenous from the blood clots or pieces of plaque from his aorta which had begun traveling to his legs. He appeared jaundiced from his failing liver and his pain was excruciating. When I recommended that he be hospitalized, he refused. Instead, as an outpatient, he attempted to continue his radiation therapy.

Finally, Mark called me. "Roy," he said sadly, "Harry is disoriented and in tremendous pain. There's no longer any question about it. He has to be hospitalized now, and Martha agrees."

"I'm sorry, Mark, it's been a long haul for everybody." I sighed sympathetically, realizing how difficult it was for Mark to lose yet another close relation. "Personally, I'm amazed he held out as long as he did. But it's time to let Harry rest. Do you think Martha will be able to handle the fact that there is nothing else to offer him except symptomatic and supportive care?"

"Deep down, I'm sure she realizes this," Mark replied wearily, "but on the surface she's still fighting."

"This won't be easy on her," I groaned and recommended an emergency admittance.

To the doctors and hospital staff, Harry's end was obviously near. Yet Martha especially and even Harry seemed unwilling to surrender their hopes despite the emergency hospitalization. At first, it surprised me that Harry, unlike most patients at end-stage wasn't appearing resigned or more accepting. Then it dawned on me. As weakened as he was, Harry was holding on for Martha's sake, as if he seemed obligated not to abandon his frightened wife.

My hospital visit with him was disturbing to me. Martha was at his bedside trying to rally the energy of her failing husband with empty hopes. She still couldn't bring herself to admit that there was nothing else to do for him. Her belief was so strong, and her desire for his recovery was so real, I could see why Harry might actually believe her. But Harry was in pain and it was time to let him go.

After I asked Martha to leave the room while I examined him, I sat down and gently asked, "Harry, do you know what's happening?"

"Yeah," he said hoarsely. "I'm going down fast."

"I agree." I said sadly. "I don't know how long we have, but we'll keep you comfortable, and maybe, just maybe there'll be a miracle." I didn't want to take the last shred of hope away from him even now. "Maybe the radiation you've received will kill the cancer..." I was finding it hard to focus on positives. "Maybe, you'll be back on your feet again—" When I made the last remark, I wanted to choke back my wrong choice of words.

What irony! Here this poor man was lying on his deathbed, with gangrene in his feet. If he didn't die before the gangrene spread, he'd probably need a bilateral amputation at the mid foot. Although a healthier man could deal with the necessary physical therapy and prosthesis to compensate for the loss of half his feet, Harry was too weak. There were just too many mountains to climb given all his circumstances, even if he had some response.

This man had very little chance for quality or quantity of life, but I couldn't leave him feeling hopeless, no matter how bad he was. Yet, I wanted him to know that if he felt it was his time, then I would trust and acknowledge his personal desire to let go of life. It was a situation which required a balance between hope and resignation, but, personally, I felt that it was a no-win situation. No matter how I handled it, someone in the family would probably be angry with me for removing his last hope—the biggest obstacle in the way of Harry's passage toward death.

"I will do everything I can for you, Harry," I assured him as I stood to leave, filling my voice with great sympathy. Gently I patted his bony hand and purposely avoided further talk about "getting on his feet." Emotions threatened to well up inside me, so instead, with the excuse that I would go find Martha, I left.

It was a disaster. I sadly reflected on the course of events which brought us to this terrible end. This is what makes people fear cancer so much. It pulls us all apart, piece by piece, response-relapse-response, pain...

When I went out into the hall, I couldn't allow myself to feel the pain. Emotionally I was like a rock now, except for some nervous energy. It was like being in an emergency situation, the emotions had shut down. I did what had to be done without distraction. Martha was standing nearby, so I approached her and tried to level with her the best I could. If she could accept the inevitable, Harry would be able to have an easier death. It was time for Martha to let him go.

Dealing with her emotions with understanding, I pointed out that her resistance to his death was not allowing him the peace to die. "He doesn't want to leave you, but he has no choice," I explained. "He's dying. Maybe we can prolong things, but the longer he lives the more pain he suffers."

Weeping, Martha finally admitted that she knew he was dying, but she didn't want him to think she had given up. "I'll try to be better with him." She said softly, while tears streamed down her face. I put my arm around her and gave her a compassionate hug. Although I still braced myself for the possibility of her anger, I assured her that I would, along with Mark and the rest of her family, help her get through this difficult time.

I left the hospital feeling tired, upset, sad, and very anxious. I had become nervous, and unsure about whether I had done the right thing for Harry. Putting myself in his position, lying in a hospital bed, I questioned whether I would want some doctor to agree with me that it was time to face my death. One part of me would be so thoroughly frightened, but another part of me would want an honest doctor to be straight with me, and yet still give me a thread of hope. This is what I hoped I accomplished with Harry that night.

Less than a week later Harry died peacefully. My relief for him was coupled with a strong sense of pain and loss. Now that he had died, my responsibility for his healthcare was over, and I could allow myself the privilege of grieving.

When I spoke with Martha soon afterwards, to express my condolences, she appeared strong and accepting. I was pleased for her. Holding tightly to Harry's memory, she was able to find the power and determination in herself and her family to continue without his physical presence.

When it came to fending off their anger, I was fortunate with Martha and Harry. Martha didn't berate me for my straightforwardness about Harry's condition. In fact, as the months passed after his death, she told me she appreciated the candid way I had answered her questions and presented the facts. It made her feel that together we had made every attempt to keep Harry alive. Finally, when he succumbed, she felt she could accept that everything that was possible had been done for him. Even for Martha, whose hold on the impossible was so strong, my honest approach was the best approach.

Often, things do not work out so nicely. Doctors have to read patient personalities in a short period of time, so there is great room for error. Some patients demand to know everything in order to feel prepared about their diseases. Others don't want to hear anything because of their tremendous fears. Most patients fall in the vast range between these two extremes, and sifting out the fearful ones from the courageous ones is no easy task. Many of these people are already emotionally and physically fragile. Learning to "tiptoe through the tulips," as one doctor referred to it, is an extremely delicate process.

Recently, a young fellow in my care, with a particularly bleak diagnosis, was faced with the horrible truth of certain death. I felt there was no way around that fact. I tried to prepare him for it by dropping sentences of a nature that didn't have a great deal of optimism, but he totally refused to hear anything negative. He only wanted reassurances from me to reaffirm his belief that he "was going to beat this thing." It was difficult. I just felt so terrible about his case, because, no matter what positive words I could offer him, his cancer was unmistakably progressing. I dreaded seeing him after a while.

In turn, he apparently didn't like me, probably because I didn't offer him the false hope he needed. Eventually he left my practice, and I felt relieved. I didn't want to be around to experience the tremendous disappointment, angst, and anger he would have toward his doctor as the disease took over. At that point, he was going to feel utterly betrayed by the physician—if he could find one—who promised him or implied success and then failed to deliver.

Over the years, I have learned the best way to deal with patients is to protect them by commission and omission from the whole truth. Although I have improved my technique, occasionally I still blunder. When I discover that, as far as the patient and family are concerned, I have said too much, too soon, and too candidly, I've apologized, admitting that I wasn't aware of how afraid they were and that I misjudged them.

However, when family members advise me to keep the patient completely in the dark, experience has shown me that the family is usually wrong. The patient needs to have some truth because, after a while, all attempts to delude or segregate the patient from that knowledge only make the patient feel more alone and untrusting. "I'm dying of cancer and no one wants to mention a word to me; no one wants to tell me what's going on. Don't leave me hanging like this!" Despite what the relatives feel, patients don't do well when they are abandoned like this.

It is always a painful process, but the doctors are responsible to get the patients, the families, and themselves through it all in the least painful way, emotionally, physically, and psychologically. The practice of medicine requires very broad shoulders.

"Bzzz." The office intercom broke into my thoughts.

"Yes?"

"Doctor Matthew Rue would like to speak to you."

"Right!" Quickly, I made the connection. Whenever I had seen or spoken to Matthew Rue lately, John Havelock, my patient from a few years ago would come to mind. I was disappointed by the way his case had ended.

After he recovered from prostate surgery and left the hospital, Havelock kept missing appointments with me. This surprised me because I thought we had developed a rapport, so I called him a number of times at his home to see what he was doing about his treatments. Always a polite man, Mr. Havelock would reply, "I'm still r'covering from da surgery," or "radiation treatment wuz helping, I cume in vhen it's dun." He was putting me off in his usual courteous manner. I realized later that he had decided he didn't want to deal with treatment any longer. Since there was no cure for him, just chemotherapy, he had decided to let his cancer run its course. It became obvious that Havelock wanted to die. It was his perogative.

Usually, I tell my patients that they are the captains of their ships. They are the ones who are ultimately responsible for themselves and their medical care. They are the ones who have to come for the follow-up visits and decide on treatment. I'm the navigator. I can tell them what the "waters" will be like to the best of my ability. I can tell them the ways that we have negotiated through this type of illness before or inform them of any newer, less-proven-but-hopeful methods available. I can enumerate the treatments, the side effects, the expected toxicity, and the potential benefits.

Still it's up to the patients to decide if they want to be treated. Although I usually volunteer it, the patients often ask me what my opinion is as far as the "best" treatment or "most effective" treatment which I often have to explain are not necessarily the same for them. Even though the most effective treatment has a higher response rate, it can be too toxic for someone who is on the verge of saying: "You know, I'm about ready to throw in the towel. I'm not sure this treatment is for me." The best treatment, then, is one that can be modified according to the patient's individual tolerance and willingness to persevere. Many times, my older patients would tell me: "I've lived seventy-five years. If I can get another year or two out of life without paying the price of terrible toxicity, let's try it, even if the response rate is lower."

"Fine," I then say, "We'll try something less toxic that has worked before for other people."

We strike a bargain. When I think they have a chance, I push very hard, using the facts, especially if I feel a patient is giving up too easily when treatment offers them the potential of positive results. For others who have less chance, I still make the same offer: "Look. You can try this. If it's too toxic for you, I can either modify it and see how you do, or you can say 'Doc, thanks but no thanks! I'll forgo this choice. I'll let you take care of me but don't treat me with chemo any more." It's their decision. I always reassure them that I will be there for them and take care of them whether or not they choose therapy in any form.

And so it was with Mr. Havelock. The bargain we finally struck with him was difficult for me to accept, because I personally felt I could have given him a little more time with effective treatment, but it was his decision. He died the way he wanted to die, without medical intervention, surrounded by family, and made comfortable with symptomatic and supportive care in the hospital.

"Hi, Roy," Matthew Rue said when I picked up the receiver. "I'd like to send a patient over to see you..."

* * * * *

Since I look upon myself as a typical oncologist when dealing with patients, I had hoped that this book would enable people to understand me, and by extension, their own doctors as well, oncologist or otherwise. Looking back, I have been distant and callused when I needed to protect myself. I have cut patients short when the time constraints wouldn't allow more. I have had to act affable even for patients and family members whose personalities were grating, obnoxious or unappreciative, while behind the straight face, the personality in me strongly disliked them. I have made errors by misjudging patients' emotional needs or have spoken too bluntly about some issues or aspects of their disease about which they were not prepared to hear. Errors, failures, anxieties, successes are all part of the physician's learning experience.

Experience and expertise often make a doctor busy, "too busy" the critics complain. As I grew wiser in the practice, I came to realize that the physicians who were good, competent, and doing well with the patients were the busy ones. These doctors got that way because they had earned their good reputations. Even though they couldn't always spend time with everybody, the good physician would spend quality time with the patients when they needed him or her most.

A number of months ago, a distinguished man came from upstate to see me personally for a consult about his prostate cancer. Initially, the man had had a brief visit with a doctor who was only moderately informed about the latest technology with regard to combination therapy. The patient left the office wondering why he paid to have a physician recommend another doctor all the way out on Long Island. What the patient didn't see was the hours his physician had spent doing his homework, researching the problem, before he referred the man to me. There were also hundreds of unseen hours which his doctor had accumulated, gaining the hands-on experience to recognize the patient's problem, to know what treatments were available as well as how to go about obtaining treatment, even if the referral was the only way.

Fortunately, sitting down with the man for a solid forty-five minutes, I was able to explain what the current technology had to offer and help him decide which choice was best for him. The man was quite appreciative of my time and praised me for such an informative and thorough consultation. True, my years of experience with prostate cancer surpassed his previous doctor's knowledge and experience in this area, but the foresight of his referring physician was part of the whole package of helping his patient come to some solution.

Rewarding relationships between physicians and patients have unfortunately become less common in practice today. One reason is because physicians are misunderstood. The public still wants us to be Marcus Welby, the ideal of yesteryear. Malpractice suits, governmental bureaucratic interference, and insurance companies have changed all that. In the last decade especially, the image of the physician has suffered greatly at the hands of bureaucrats and litigators. By blaming doctors for the skyrocketing costs of medicine, they are stoking this antiphysician fire, fanning the flames of discontent and distrust so that both doctors and patients eye each other in this light as adversaries, not as allies.

It's true. Medicine is very expensive for many reasons. Scientists, researchers, drug companies, and physicians are all making great strides in improving the technology, but the cost of progress is quite high. As a result of such advances, however, the United States is still one of the best places in the world to receive treatment. Dignitaries as well as hardship cases from countries around the globe come here for help.

Still, there is no disputing that medical costs should be curbed. Believing that they could help the ailing health care system, federal and state governments and insurance companies have stepped in over the years. Instead of making matters better, they have made matters worse. Without thoroughly considering the consequences on medical care, Congressmen are legislating cut after cut. The D.R.G.s deprive some patients of adequate hospitalization with inappropriate limitations. Fee schedules and freezes deny some doctors as well as patients their realistic reimbursements for actual costs of drugs and treatments, while others, of course, are overpaid. Imbalances are upsetting the system. The list of problems continues.

Over the last several years, the marketplace has become much more competitive. This has been, in part, due to the Reagan administration's belief that increased competition would serve to lower costs. For a period of time, there was a push to increase the number of doctors in the country. This had two unexpected and adverse effects.

First, the concept that larger numbers of physicians meant decreased financial rewards served to reduce medical school applications. For a period of time, medical schools had to decrease their standards in order to fill their classes. Had this continued for any period of time, the future of quality of healthcare in this country would certainly have been threatened.

Second, where at one time an oncologist, for example, would see hundreds of patients in a year, gaining greater knowledge in the craft of treating patients, the frightening prediction—fast becoming true—was that during the early 1990s, the average oncologist across the country would see only a hundred new patients a year. This decrease in patient load gives the doctor less exposure to a variety of cancers. Less experience with treating patients for the variety of cancers means a lower level of diagnostic and therapeutic acumen, all resulting in detrimental effects on actual patient care. Simply put, when the patient load goes too low, doctors lose their expertise.

More recently, the march toward managed care has led to a series of new problems. The bureaucratic nightmare of the approval process has led to a greater financial cost and increased frustrations in caring for patients. Frequently busy phone lines, inappropriately trained and unqualified triage staff, among other serious defects, lead to greater discontent among healthcare providers.

Physicians who at first resisted the reduced fee structure of managed care organizations have seen their patients defect to other doctors who signed up with these plans. Many doctors who decided later to join were told the plans were full. We are now seeing a new trend: patients ending up with doctors who are "listed in the book," rather than because of word-of-mouth recommendations or reputation. This has tended to diminish the doctor-patient relationship.

As these HMO's and managed care organizations continue to control the medical care system, I see a further dehumanization of healthcare delivery taking place. "Capitated plans," that is, plans that set a fixed fee per head, especially lead to this. The more patients a doctor has in his practice, each for a fixed reimbursement per month, the less incentive a doctor has to spend quality time with such patients. Under this system, neither the patient nor the doctor will be satisfied.

There is no denying that we are all obligated to help lower the national deficit, but there is an undue burden on the hospital and medical care system, not just the doctors, imposed by our bureaucratic system. At every hospital I visit, I see signs that they are hurting. Almost every hospital on Long Island is operating in the red. This deleterious condition is nationwide.

Patient care has to be affected when a hospital operates in the red. They're not going to hire the same number of nurses. The patients anguishing in the bed are not going to get their narcotics or medicines as fast as they would if the hospital was operating in the black. It's just a fact of life: cut the funds, and the level of service goes down.

Whereas many people don't want to pay the price, ultimately, the price they will be paying is inferior health care. Already, interference from government, insurance companies, and other industries makes medicine more and more difficult to practice and more cold and commercial in the delivery of care. Having more doctors hasn't lowered the cost of the drugs and medical supplies.

It decreases the expertise of the physicians themselves. Less funding to hospitals has frequently made patient care less than adequate. Finally, when the image of the physician becomes tarnished, and trust is taken away, the doctor-patient relationship deteriorates. Caregiving becomes an unrewarding task. This further contributes to the deterioration of the entire medical care system. These problems and others growing from them face us if medicine continues in the direction in which it has been going over the last ten years.

Obviously, something has to change if we want the United States to maintain its excellence in healthcare today as well as in the medical practice of tomorrow. That change should be in attitude.

Save the system! It's not a bad system. It only needs to be modified. Yes, it's a costly one, but there are ways to save it. The most important thing to every human being is their health. It should be number one in terms of the amount of money allocated. I'm not saying spend more, all I'm saying is re-appropriate what is already there. For example, two areas in which many more of our healthcare dollars are being spent than should be are on liability insurance (malpractice) and administration costs. Such excess must be reduced, but trim the fat away with a scalpel, not with an ax.

So far, I've defined many of the problems, difficulties, sadness, and tragedies that beset doctors in a practice today. Yet look at what we've accomplished. I see it every time I find another successful protocol, another prolonged remission, another cure. It must not stop because the priorities of the country are confused. Without the combined forces of federal and state governments, insurance companies, and medical professionals listening to each other, working together, this system will not succeed, and the hopes of millions of people suffering from cancer, the number two killer in the United States, will not come to pass.

What are tomorrow's miracles? I look toward the future with tremendous excitement, optimism, and hope. Even the failures of today will lead us to the successes of tomorrow. Already technological advances in the discovery at the molecular level of genes have more precisely identified oncogenes—those genes that are responsible for causing cancer—and proto-oncogenes—cancer repressors. New biologicals such as G-CSF and GM-CSF allow us to use more toxic drugs with less harm to normal tissue such as bone marrow. New anti-emetics drugs successfully control nausea and vomiting from chemotherapy in over eighty percent of patients. This means that while treatments will become more intense, patients will be better able to tolerate the high doses. Advances in diagnoses and staging with the use of PET, MRI, and CAT scans have already improved our ability to see into the human body in noninvasive ways. With all the diagnostic and technological tools, we are just beginning to catch cancer in its infancy. Assisted by the computerized compilation of data—a task that at one time would have been impossible—medicine of today will provide better clues for researchers of tomorrow. Indeed, we are in very exciting times.

So what will the future hold for medicine? I see it clearly. More advances in cancer research bring us more answers. More answers lead us to more successes, and more successes allow more people to get on with their lives. When treatments get better, disappointments and failures become fewer. As the aforementioned navigator on board my patients' ships, I will be able to steer through an otherwise difficult ocean, with improved chances of achieving a smoother and more successful journey to port. These visions can only come true if people have an active input in the current political process that will reshape our healthcare system.

Finally, if the general public could comprehend that inherent to and at the root of today's healthcare delivery, even at the highest technologic level, is the interpersonal patient-physician relationship, the best of the medical system can be saved for tomorrow. If this bond can be strengthened, if the doctor and patient can respect each other, then both will ultimately benefit. I hope my story has shown that doctors are people reacting like people to these trying times—we feel, we hurt, we cry, we laugh, we are human in every way, just like our patients and their families—and that we are trying our best to be there for our patients. Perhaps then, people will understand what it is like to be the person on the other side of the stethoscope.

Co-author: Linda A. Mittiga, writer

As a long-time cancer survivor and Dr. Berger's patient, co-writer Linda A. Mittiga knew first-hand the ordeals of fighting cancer. Cancer suspended her career as a book editor for a large publishing firm in New York City. It disrupted her personal plans as a newlywed contemplating the possibility of children with her husband. Cancer, an interruption that threatened to take her life, drastically changed her life by putting her in touch with Doctor Roy Berger who found the treatment for her cure. The bond that developed between doctor and patient taught Linda Mittiga about the disease from a new perspective—that of a cancer doctor. Through their collaboration on this book, she learned that, for oncologists, cancer is not an interruption; it is a constant way of life, filled with fateful cruelty and human compassion.

Cured now for many years, Linda Mittiga is a writer/computer graphics designer for a public relations firm on Long Island where she and her husband live with their three, wonderful children. Aware of her good fortune, she appreciates God's gift of each new day in good health.

Printed in the United States
62168LVS00004B/322-381